John Brockman is a writer, agent and publisher of the 'Third Culture' website www.edge.org, the forum for leading scientists and thinkers to share their research with the general public. He is the author of *The Third Culture* and has edited several previous anthologies including *The Next Fifty Years* and *What We Believe But Cannot Prove.*

BOOKS BY JOHN BROCKMAN

As author:
By the Late John Brockman
37
Afterwords
The Third Culture: Beyond the Scientific Revolution
Digerati

As editor:
About Bateson
Speculations
Doing Science
Ways of Knowing
Creativity
The Greatest Inventions of the Past 2,000 Years
The Next Fifty Years
The New Humanists
Curious Minds
What We Believe But Cannot Prove
My Einstein
Intelligent Thought

As coeditor:
How Things Are

WHAT IS YOUR DANGEROUS IDEA?

Today's Leading Thinkers on the Unthinkable

edited by JOHN BROCKMAN

POCKET
BOOKS

LONDON • SYDNEY • NEW YORK • TORONTO

First published in Great Britain by Simon & Schuster UK Ltd, 2006
This edition first published by Pocket Books, 2007
An imprint of Simon & Schuster UK Ltd
A CBS COMPANY

1 3 5 7 9 10 8 6 4 2

Simon & Schuster UK Ltd
Africa House
64–78 Kingsway
London WC2B 6AH

www.simonsays.co.uk

Simon & Schuster Australia
Sydney

A CIP catalogue record for this book
is available from the British Library.

ISBN-13: 978-1-4165-2685-8
ISBN-10: 1-4165-2685-4

Typeset in Garamond by M Rules
Printed and bound in Great Britain by
Cox & Wyman Ltd, Reading, Berks

To Stewart Brand, George Dyson and Kevin Kelly
for their advice and support during the first
ten years of the *Edge* conversation

CONTENTS

ACKNOWLEDGMENTS

I wish to thank my U.S. publisher, John Williams of HarperCollins, and my U.K. publisher, Andrew Gordon of Simon & Schuster, for their encouragement.

I am also indebted to my agent, Max Brockman, who recognized the potential for this book, and to Sara Lippincott for her thoughtful and meticulous editing.

PREFACE : THE *EDGE* QUESTION

In 1991, I suggested the idea of a third culture, which 'consists of those scientists and other thinkers in the empirical world who, through their work and expository writing, are taking the place of the traditional intellectual in rendering visible the deeper meanings of our lives, redefining who and what we are.' By 1997, the growth of the Internet had allowed implementation of a home for the third culture on the Web, on a site named *Edge* (www.edge.org).

Edge is a celebration of the ideas of the third culture, an exhibition of this new community of intellectuals in action. They present their work, their ideas, and comment about the work and ideas of third culture thinkers. They do so with the understanding that they are to be challenged. What emerges is rigorous discussion concerning crucial issues of the digital age in a highly charged atmosphere where 'thinking smart' prevails over the anesthesiology of wisdom.

The ideas presented on *Edge* are speculative; they represent the frontiers in such areas as evolutionary biology, genetics, computer science, neurophysiology, psychology, and physics. Some of the fundamental questions posed are: Where did the universe come from? Where did life come from? Where did the mind come from? Emerging out of the third culture is a new natural philosophy, new ways of understanding physical systems, new ways of thinking that call into question many of our basic assumptions of who we are, of what it means to be human.

An annual feature of *Edge* is The World Question Center, which was introduced in 1971 as a conceptual art project by my friend and collaborator the late artist James Lee Byars. His plan was to gather the hundred most brilliant minds in the world together in a room, lock them in, and 'have them ask each other the questions they were asking themselves.' The result was to be a synthesis of all thought. Between idea and execution, however, are many pitfalls. Byars identified his hundred most brilliant minds, called each of them, and asked them what questions they were asking themselves. The result: seventy people hung up on him.

But by 1997, the Internet and e-mail had allowed for a serious implementation of Byars' grand design, and this resulted in launching *Edge.* For each of the anniversary editions of *Edge,* I have used the interrogative myself and asked contributors for their responses to a question that comes to me, or to one of my correspondents, in the middle of the night. The 2006 *Edge* Question was suggested by the psychologist Steven Pinker:

The history of science is replete with discoveries that were considered socially, morally, or emotionally dangerous in their time; the Copernican and Darwinian revolutions are the most obvious. What is your dangerous idea? An idea you think about (not necessarily one you originated) that is dangerous not because it is assumed to be false, but because it might be true?

The 2005 *Edge* Question 'What do you believe is true, even though you cannot prove it?', was an eye-opener (BBC4 Radio characterized it as 'fantastically stimulating . . . the crack cocaine of the thinking world'). It is hoped that this edition of responses to the 2006 *Edge* Question will be equally as dangerous.

John Brockman
Publisher and Editor, *Edge*

INTRODUCTION

Do women, on average, have a different profile of aptitudes and emotions than men? Were the events in the Bible fictitious — not just the miracles, but those involving kings and empires? Has the state of the environment improved in the last fifty years? Do most victims of sexual abuse suffer no lifelong damage? Did Native Americans engage in genocide and despoil the landscape? Do men have an innate tendency to rape? Did the crime rate go down in the 1990s because two decades earlier poor women aborted children who would have been prone to violence? Are suicide terrorists well educated, mentally healthy, and morally driven? Are Ashkenazi Jews, on average, smarter than gentiles because their ancestors were selected for the shrewdness needed in money lending? Would the incidence of rape go down if prostitution were legalized? Do African American men have higher levels of testosterone, on average, than white men? Is morality just a product of the evolution of our brains, with no inherent reality? Would society be better off if heroin and cocaine were legalized? Is homosexuality the symptom of an infectious disease? Would it be consistent with our moral principles to give parents the option of euthanizing newborns with birth defects that would consign them to a life of pain and disability? Do parents have any effect on the character or intelligence of their children? Have religions killed a greater proportion of people than Nazism? Would damage from terrorism be reduced if the police could torture suspects in special circumstances? Would Africa have a better chance of rising out of poverty if it hosted more polluting industries or accepted Europe's nuclear waste? Is the average intelligence of Western nations declining because duller people are having more children than smarter people? Would unwanted children be better off if there were a market in adoption rights, with babies going to the highest bidder? Would lives be saved if we instituted a free market in organs for transplantation? Should people have the right to clone themselves, or enhance the genetic traits of their children?

Perhaps you can feel your blood pressure rise as you read these questions. Perhaps you are appalled that people can so much as *think* such things. Perhaps you think less of *me* for bringing them up. These are dangerous ideas – ideas that are denounced not because they are self-evidently false, nor because they advocate harmful action, but because they are thought to corrode the prevailing moral order.

By 'dangerous ideas' I don't have in mind harmful technologies, like those behind weapons of mass destruction, or evil ideologies, like those of racist, fascist, or other fanatical cults. I have in mind statements of fact or policy that are defended with evidence and argument by serious scientists and thinkers but which are felt to challenge the collective decency of an age. The ideas in the first paragraph, and the moral panic that each one of them has incited during the past quarter century, are examples. Writers who have raised ideas like these have been vilified, censored, fired, threatened, and in some cases physically assaulted.

Every era has its dangerous ideas. For millennia, the monotheistic religions have persecuted countless heresies, together with nuisances from science such as geocentrism, biblical archeology, and the theory of evolution. We can be thankful that the punishments have changed from torture and mutilation to the canceling of grants and the writing of vituperative reviews. But intellectual intimidation, whether by sword or by pen, inevitably shapes the ideas that are taken seriously in a given era, and the rear-view mirror of history presents us with a warning. Time and again people have invested factual claims with ethical implications that today look ludicrous. The fear that the structure of our solar system has grave moral consequences is a venerable example, and the foisting of 'Intelligent Design' on biology students is a contemporary one. These travesties should lead us to ask whether the contemporary intellectual mainstream might be entertaining similar moral delusions. Are we liable to be enraged by our own infidels and heretics whom history may some day vindicate?

I suggested to John Brockman that he devote his annual *Edge* question to dangerous ideas because I believe that they are likely to confront us at an increasing rate and that we are ill-equipped to deal with them. When done right, science (together with other truth-seeking institutions, such as history and journalism) characterizes the world as it is, without regard to whose feelings get hurt. Science in particular has always been a source of heresy, and today the galloping advances in touchy areas like genetics, evolution, and the environment sciences are bound to throw unsettling possibilities at us. Moreover, the rise of globalization and the Internet are allowing heretics to find one another and work around the barriers of traditional media and academic journals. I also suspect that a change in generational sensibilities will hasten the process. The term 'political correctness' captures the 1960s conception of moral rectitude that we baby boomers brought with us as we took over academia, journalism, and government. In my experience, today's students – black and white, male and female – are bewildered by the idea, common among their parents, that certain scientific opinions are immoral or certain questions too hot to handle.

What makes an idea 'dangerous'? One factor is an imaginable train of events in which acceptance of the idea could lead to an outcome that only recently has been recognized as harmful. In religious societies, the fear is that if people ever stopped believing in the literal truth of the Bible they would also stop believing in the authority of its moral commandments. That is, if today people dismiss the part about God creating the earth in six days, tomorrow they'll dismiss the part about 'Thou shalt not kill.' In progressive circles, the fear is that if people ever were to acknowledge any differences between races, sexes, or individuals, they would feel justified in discrimination or oppression. Other dangerous ideas set off fears that people will neglect or abuse their children, become indifferent to the environment, devalue human life, accept violence, and prematurely resign themselves to social problems that could be solved with sufficient commitment and optimism.

All these outcomes, needless to say, would be deplorable. But none of them actually follows from the supposedly dangerous idea. Even if it turns out, for instance, that groups of people are different in their averages, the overlap is certainly so great that it would be irrational and unfair to discriminate against individuals on that basis. Likewise, even if it turns out that parents don't have the power to shape their children's personalities, it would be wrong on grounds of simple human decency to abuse or neglect one's children. And if currently popular ideas about how to improve the environment are shown to be ineffective, it only highlights the need to know what *would* be effective.

Another contributor to the perception of dangerousness is the intellectual blinkers that humans tend to don when they split into factions. People have a nasty habit of clustering in coalitions, professing certain beliefs as badges of their commitment to the coalition and treating rival coalitions as intellectually unfit and morally depraved. Debates between members of the coalitions can make things even worse, because when the other side fails to capitulate to one's devastating arguments, it only proves they are immune to reason. In this regard, it's disconcerting to see the two institutions that ought to have the greatest stake in ascertaining the truth – academia and government – often blinkered by morally tinged ideologies. One ideology is that humans are blank slates and that social problems can be handled only through government programs that especially redress the perfidy of European males. Its opposite number is that morality inheres in patriotism and Christian faith and that social problems may be handled only by government policies that punish the sins of individual evildoers. New ideas, nuanced ideas, hybrid ideas – and sometimes dangerous ideas – often have trouble getting a hearing against these group-bonding convictions.

The conviction that honest opinions can be dangerous may even arise from a feature of human nature. Philip Tetlock and Alan Fiske have argued that certain human relationships are constituted on a basis of unshakeable convictions. We love our children and parents, are faithful to our spouses, stand by our friends, contribute to our communities, and are loyal to our coalitions not because we continually question and evaluate the merits of these commitments but because we feel them in our bones. A person who spends too much time pondering whether logic and fact really justify a commitment to one of these relationships is seen as just not 'getting it.' Decent people don't carefully weigh the advantages and disadvantages of selling their children or selling out their friends or their spouses or their colleagues or their country. They reject these possibilities outright; they 'don't go there.' So the taboo on questioning sacred values makes sense in the context of personal relationships. It makes far less sense in the context of discovering how the world works or running a country.

Should we treat some ideas as dangerous? Let's exclude outright lies, deceptive propaganda, incendiary conspiracy theories from malevolent crackpots, and technological recipes for wanton destruction. Consider only ideas about the truth of empirical claims or the effectiveness of policies that, if they turned out to be true, would require a significant rethinking of our moral sensibilities. And consider ideas that, if they turn out to be false, could lead to harm if people believed them to be true. In either case, we don't know whether they are true or false *a priori*, so only by examining and debating them can we find out. Finally, let's assume that we're not talking about burning people at the stake or cutting out their tongues but about discouraging their research and giving their ideas as little publicity as possible.

There is a good case for exploring all ideas relevant to our current concerns, no matter where they lead. The very act of engaging in rational discourse presupposes a commitment to evaluating ideas on their intellectual warrant alone. Otherwise how could one even make the case that dangerous ideas should be discouraged, in the face of someone else arguing (as Dan Gilbert does in this volume) that the idea of discouraging ideas is itself morally dangerous? Should proponents of keeping dangerous ideas private then be forced to keep *that* idea private, because their opponents deem *it* to be dangerous? If not, why should the proponents' judgment about dangerousness and nondangerousness be granted a privilege they deny to others? The idea that ideas should be discouraged *a priori* is inherently self-refuting. Indeed, it is the ultimate arrogance, as it assumes that one can be so certain about the goodness and truth of one's own ideas that one is entitled to discourage other people's opinions from even being examined.

Also, it's hard to imagine any aspect of public life where ignorance or delusion is better than an awareness of the truth, even an unpleasant one. Only children and madmen engage in 'magical thinking,' the fallacy that good things can come true by believing in them or bad things will disappear by ignoring them or wishing them away. Rational adults want to know the truth, because any action based on false premises will not have the effects they desire. Worse, logicians tell us that a system of ideas containing a contradiction can be used to deduce any statement whatsoever, no matter how absurd. Since ideas are connected to other ideas, sometimes in circuitous and unpredictable ways, choosing to believe something that may not be true, or even maintaining walls of ignorance around some topic, can corrupt all of intellectual life, proliferating error far and wide. In our everyday lives, would we want to be lied to, or kept in the dark by paternalistic 'protectors,' when it comes to our health or finances or even the weather? In public life, imagine someone saying that we should not do research into global warming or energy shortages because if it found that they were serious the consequences for the economy would be extremely unpleasant. Today's leaders who tacitly take this position are rightly condemned by intellectually responsible people. But why should other unpleasant ideas be treated differently?

There is another argument against treating ideas as dangerous. Many of our moral and political policies are designed to pre-empt what we know to be the worst features of human nature. The checks and balances in a democracy, for instance, were invented in explicit recognition of the fact that human leaders will always be tempted to arrogate power to themselves. Likewise, our sensitivity to racism comes from an awareness that groups of humans, left to their own devices, are apt to discriminate and oppress other groups, often in ugly ways. History also tells us that a desire to enforce dogma and suppress heretics is a recurring human weakness, one that has led to recurring waves of gruesome oppression and violence. A recognition that there is a bit of Torquemada in everyone should make us wary of any attempt to enforce a consensus or demonize those who challenge it.

'Sunlight is the best disinfectant,' according to Justice Louis Brandeis's famous case for freedom of thought and expression. If an idea really is false, only by examining it openly can we determine that it is false. At that point we will be in a better position to convince others that it is false than if we had let it fester in private, since our very avoidance of the issue serves as a tacit acknowledgment that it may be true. And if an idea is *true*, we had better accommodate our moral sensibilities to it, since no good can come from sanctifying a delusion. This might even be easier than the ideaphobes fear. The moral order did not collapse when the earth was shown not to be at the center of the solar system, and so it will survive other revisions of our understanding of how the world works.

In the best Talmudic tradition of arguing a position as forcefully as possible and then switching sides, let me now present the case for *discouraging* certain lines of intellectual inquiry. Two of the contributors to this volume (Gopnik and Hillis) offer as their 'dangerous idea' the exact opposite of Gilbert's: They say that it's a dangerous idea for thinkers to air their dangerous ideas. How might such an argument play out?

First, one can remind people that we are all responsible for the foreseeable consequences of our actions, and that includes the consequences of our public statements. Freedom of inquiry may be an important value, according to this argument, but it is not an *absolute* value, one that overrides all others. We know that the world is full of malevolent and callous people who will use any pretext to justify their bigotry or destructiveness. We must expect that they will seize on the broaching of a topic that seems in sympathy with their beliefs as a vindication of their agenda.

Not only can the imprimatur of scientific debate add legitimacy to toxic ideas, but the mere act of making an idea common knowledge can change its effects. Individuals, for instance, may harbor a private opinion on differences between genders or among ethnic groups but keep it to themselves because of its opprobrium. But once the opinion is aired in public, they may be emboldened to act on their prejudice – not just because it has been publicly ratified but because they must anticipate that *everyone else* will act on the information. Some people, for example, might discriminate against the members of an ethnic group despite having no pejorative opinion about them, in the expectation that their customers or colleagues *will* have such opinions and that defying them would be costly. And then there are the effects of these debates on the confidence of the members of the stigmatized groups themselves.

Of course, academics can warn against these abuses, but the qualifications and nitpicking they do for a living may not catch up with the simpler formulations that run on swifter legs. Even if they did, their qualifications might be lost on the masses. We shouldn't count on ordinary people to engage in the clear thinking – some would say the hair-splitting – that would be needed to accept a dangerous idea but not its terrible consequence. Our overriding precept, in intellectual life as in medicine, should be 'First, do no harm.'

We must be especially suspicious when the danger in a dangerous idea is to someone other than its advocate. Scientists, scholars, and writers are members of a privileged elite. They may have an interest in promulgating ideas that justify their privileges, that blame or make light of society's victims, or that earn them attention for cleverness and iconoclasm. Even if one has little sympathy for the cynical Marxist argument that ideas are always advanced to serve the interest of the ruling class, the ordinary skepticism of a tough-minded intellectual (the mindset that leads us to blind review, open debate, and statements of possible conflicts of interest) should make one wary of 'dangerous' hypotheses that are no skin off the nose of their hypothesizers.

But don't the demands of rationality always compel us to seek the complete truth? Not necessarily. Rational agents often choose to be ignorant. They may decide not to be in a position where they can receive a threat or be exposed to a sensitive secret. They may choose to avoid being asked an incriminating question, where one answer is damaging, another is dishonest, and a failure to answer is grounds for the questioner to assume the worst (hence the Fifth Amendment protection against being forced to testify against oneself). Scientists test drugs in double-blind studies in which they keep themselves from knowing who got the drug and who got the placebo, and they referee manuscripts anonymously for the same reason. Many people rationally choose not to know the gender of their unborn child, or whether they carry a gene for Huntington's disease, or whether their nominal father is genetically related to them. Perhaps a similar logic would call for keeping socially harmful information out of the public sphere.

As for restrictions on inquiry, every scientist already lives with them. They accede, for example, to the decisions of committees for the protection of human subjects and to policies on the confidentiality of personal information. In 1975 biologists imposed a moratorium on research on recombinant DNA pending the development of safeguards against the release of dangerous microorganisms. The notion that intellectuals have *carte blanche* in conducting their inquiry is a myth.

Though I am more sympathetic to the argument that important ideas be aired than to the argument that they should sometimes be suppressed, I think it is a debate we need to have. Whether we like it or not, science has a habit of turning up discomfiting thoughts, and the Internet has a habit of blowing their cover. Tragically, there are few signs that the debates will happen in the place where we might most expect it: academia. Though academics owe the extraordinary perquisite of tenure to the ideal of encouraging free inquiry and the evaluation of unpopular ideas, all too often academics are the first to try to quash them. The most famous recent example is the outburst of fury and disinformation that resulted when Harvard president

Lawrence Summers gave a measured analysis of the multiple causes of women's underrepresentation in science and math departments in elite universities and tentatively broached the possibility that discrimination and hidden barriers were not the only cause. But intolerance of unpopular ideas among academics is an old story. Books like Morton Hunt's *The New Know-Nothings* and Alan Kors and Harvey Silverglate's *The Shadow University* have depressingly shown that universities cannot be counted on to defend the rights of their own heretics and that it's often the court system or the press that has to drag them into policies of tolerance. In government, the intolerance is even more frightening, because the ideas considered there are not just matters of intellectual sport but have immediate and sweeping consequences. Chris Mooney, in *The Republican War on Science*, joins Hunt in showing how corrupt and demagogic legislators are increasingly stifling research findings they find inconvenient to their interests.

The essays in the present volume offer a startling variety of stimulating thoughts. Some are frankly speculative, others are ideas about an unrecognized danger, and many are versions of Copernicus's original dangerous idea – that we are not the center of the universe, literally or metaphorically. Whether you agree or disagree, are shocked or blasé, I hope that these essays provoke you to ponder what makes ideas dangerous and what we should do about them.

Steven Pinker

The history of science is replete with discoveries that were considered socially, morally, or emotionally dangerous in their time; the Copernican and Darwinian revolutions are the most obvious. What is your dangerous idea? An idea you think about (not necessarily one you originated) that is dangerous not because it is assumed to be false, but because it might be true?

We Have No Souls

JOHN HORGAN

John Horgan is the director of the Center for Science Writings at Stevens Institute of Technology and the author, most recently, of *Rational Mysticism: Spirituality Meets Science in the Search for Enlightenment*.

This year's *Edge* question makes me wonder: Which ideas pose a greater potential danger? False ones or true ones? Illusions or the lack thereof? As a believer in and lover of science, I certainly hope that the truth will set us free, and save us, but sometimes I'm not so sure.

The dangerous (probably true) idea I'd like to dwell on is that we humans have no souls. The soul is that core of us that supposedly transcends, and even persists beyond, our physicality, lending us a fundamental autonomy, privacy, and dignity. In his 1994 book *The Astonishing Hypothesis: The Scientific Search for the Soul*, the late, great Francis Crick argued that the soul is an illusion perpetuated, like Tinkerbell, only by our belief in it. Crick opened his book with this manifesto: "'You,' your joys and your sorrows, your memories and your ambitions, your sense of personal identity and free will, are in fact no more than the behavior of a vast assembly of nerve cells and their associated molecules.' Note the quotation marks around 'You.' The subtitle of Crick's book was almost comically ironic, since he was clearly trying not to find the soul but to crush it out of existence.

I once told Crick that 'The Depressing Hypothesis' would have been a more accurate title for his book, since he was, after all, just reiterating the basic, materialist assumption of modern neurobiology and, more broadly, all of science. Until recently, it was easy to dismiss this assumption as moot, because brain researchers had made so little progress in tracing cognition to specific neural processes. Even self-proclaimed materialists, who intellectually accept the idea that we are just meat machines, could harbor a secret sentimental belief in a soul of the gaps. But recently the gaps have been closing, as neuroscientists – egged on by Crick in the last two decades of his life – have begun unraveling the so-called neural code, the software that transforms electrochemical pulses in the brain into perceptions, memories, decisions, emotions, and other constituents of consciousness.

I've argued elsewhere that the neural code may turn out to be so complex that it will never be fully deciphered. But sixty years ago some biologists feared the genetic code was too complex to crack. Then in 1953 Crick and James Watson unraveled the structure of DNA, and researchers quickly established that the double helix mediates an astonishingly simple genetic code governing the heredity of all organisms. Science's success in deciphering the genetic code, which has culminated in the Human Genome Project, has been widely acclaimed – and with good reason, because knowledge of our genetic makeup could allow us to reshape our innate nature. A solution to the neural code could give us much greater, more direct control over ourselves than mere genetic manipulation.

Will we be liberated or enslaved by this knowledge? Officials in the Pentagon, the major funder of neural code research, have openly broached the prospect of cyborg warriors who can be remotely controlled via brain implants, like the assassin in the recent remake of *The Manchurian Candidate*. On the other hand, a cult-like group of self-described 'wireheads' looks forward to the day when implants allow us to create our own realities and achieve ecstasy on demand.

Either way, when our minds can be programmed like personal computers, then perhaps we will finally abandon the belief that we have immortal, inviolable souls – unless, of course, we program ourselves to believe.

The Rejection of the Soul

PAUL BLOOM

Paul Bloom is a psychologist at Yale University and the author of *Descartes' Baby: How the Science of Child Development Explains What Makes Us Human*.

I am not concerned here with the radical claim that personal identity, free will, and consciousness do not exist. Regardless of its merit, this position is so intuitively outlandish that nobody but a philosopher could take it seriously, and so it is unlikely to have any real-world implications, dangerous or otherwise.

Instead I am interested in the milder position that mental life has a purely material basis. The dangerous idea, then, is that Cartesian dualism is false. If what you mean by 'soul' is something immaterial and immortal, something that exists independently of the brain, then souls do not exist. This is old hat for most psychologists and philosophers, the stuff of introductory lectures. But the rejection of the immaterial soul is unintuitive, unpopular, and, for some people, downright repulsive.

In the journal *First Things*, Patrick Lee and Robert P. George outline some worries from a religious perspective:

If science did show that all human acts, including

conceptual thought and free choice, are just brain processes, . . . it would mean that the difference between human beings and other animals is only superficial – a difference of degree rather than a difference in kind; it would mean that human beings lack any special dignity worthy of special respect. Thus, it would undermine the norms that forbid killing and eating human beings as we kill and eat chickens, or enslaving them and treating them as beasts of burden as we do horses or oxen.

The conclusions don't follow. Even if there are no souls, humans might differ from nonhuman animals in some other way, perhaps in their capacity for language or abstract reasoning or emotional suffering. And even if there were no difference, it would hardly give us license to do terrible things to human beings. Instead, as the philosopher Peter Singer and others have argued, it should make us kinder to nonhuman animals. If a chimpanzee turned out to possess the intelligence and emotions of a human child, for instance, most of us would agree that it would be wrong to eat, kill, or enslave it.

Still, Lee and George are right to worry that giving up on the soul means giving up on an *a priori* distinction between humans and other creatures, something which has very real consequences. It would affect as well how we think about stem-cell research and abortion, euthanasia, cloning, and cosmetic psychopharmacology. It would have substantial implications in the legal realm: A belief in immaterial souls has led otherwise sophisticated commentators to defend a distinction between actions that we do and actions that our brains do. We are responsible only for the former, motivating the excuse that the cognitive neuroscientist Michael Gazzaniga has called 'My brain made me do it.' It has been proposed, for instance, that if a pedophile's brain shows a certain pattern of activation while contemplating sex with a child, he should not be held fully responsible for his actions. When you give up on the soul and accept that all actions correspond to brain activity, this sort of reasoning goes out the window.

The rejection of souls is more dangerous than evolution by natural selection. The battle between evolution and creationism is important for many reasons; it is where science takes a stand against superstition. But, like the origin of the universe, the origin of the species is an issue of great intellectual importance and little practical relevance. If everyone were to become a sophisticated Darwinian, our everyday lives would change very little. In contrast, the widespread rejection of the soul would have profound moral and legal consequences. It would also require people to rethink what happens when they die, and give up the idea (held by some 90 percent of Americans) that their souls will survive the death of their bodies and ascend to heaven. It is hard to get more dangerous than that.

The Evolution of Evil

DAVID BUSS

David Buss is a psychologist at the University of Texas, Austin, and the author of *The Murderer Next Door: Why the Mind is Designed to Kill*.

When most people think of torturers, stalkers, robbers, rapists, and murderers, they imagine crazed, drooling monsters, with maniacal Charles Manson-like eyes. The calm, normal-looking image staring back at you from the bathroom mirror reflects a truer representation. My dangerous idea is that all of us contain, within our large brains, adaptations whose functions are to commit despicable atrocities against our fellow humans – atrocities that most would label evil.

The unfortunate fact is that killing has proved to be an effective solution to an array of adaptive problems in the ruthless evolutionary games of survival and reproductive competition: defending oneself against injury, rape, or death; protecting one's children; eliminating an antagonist; acquiring a rival's resources; securing sexual access to a competitor's mate; preventing an interloper from appropriating one's own mate; protecting vital resources needed for reproduction.

The idea that evil has evolved is dangerous on several counts. If our brains contain psychological circuits that can trigger murder, genocide, and other forms of malevolence, then perhaps we can't hold those who commit carnage responsible. ('It's not my client's fault, your honor. His evolved homicidal adaptations made him do it.') Understanding causality, however, does not exonerate murderers, whether the tributaries lead back to human evolutionary history or to modern exposure to alcoholic mothers, violent fathers, or the ills of bullying, poverty, drugs, or computer games. It would be dangerous if the theory of the evolved murderous mind were misused to let killers go free.

The idea of the evolution of evil is dangerous for a more disconcerting reason. We like to believe that evil can be objectively located in a particular set of evil deeds, or within the subset people who perpetrate horrors on others, regardless of the perspective of the perpetrator or victim. That is not the case. The perspective of the perpetrator and victim differ profoundly. Many view killing a member of one's ingroup, for example, to be evil but take a different view of killing those in the outgroup. Some people point to the biblical commandment 'Thou shalt not kill' as an absolute. Closer biblical inspection reveals that this injunction applied only to murder within one's group.

Conflict with terrorists provides a modern example. Osama bin Laden declared, 'The ruling to kill the Americans and their allies – civilians and military – is an individual duty for every Muslim who can do it in any country in which it is possible to do it.' What is evil from the perspective of an American who is a potential victim is an act of responsibility and higher moral good from the terrorist's perspective. Similarly, when President Bush identified an 'axis of evil,' he rendered it moral for Americans to kill those falling under that axis – a judgment undoubtedly considered evil by those whose lives have become imperiled.

At a rough approximation, we view as evil those people who inflict massive evolutionary fitness costs on us, our families, or our allies. No one summarized these fitness costs better than the feared conqueror Genghis Khan (1167–1227): 'The greatest pleasure is to vanquish your enemies, to chase them before you, to rob them of their wealth, to see their near and dear bathed in tears, to ride their horses and sleep on the bellies of their wives and daughters.'

We can be sure that the families of the victims of Genghis Khan saw him as evil. We can be just as sure that his many sons, whose harems he filled with women of the conquered groups, saw him as a venerated benefactor. In modern times, we react with horror at Mr Khan describing the deep psychological satisfaction he gained from inflicting fitness costs on victims while purloining fitness fruits for himself. But it is sobering to realize that perhaps half a percent of the world's population today are his descendants.

On reflection, the dangerous idea may not be that murder historically has been advantageous to the reproductive success of killers, nor that we all house homicidal circuits within our brains, nor even that all of us are lineal descendants of ancestors who murdered. The danger comes from people who refuse to recognize that there are dark sides of human nature that cannot be wished away by attributing them to the modern ills of culture, poverty, pathology, or exposure to media violence. The danger comes from failing to gaze into the mirror and come to grips with the capacity for evil in all of us.

The Differences Between Humans and Nonhumans Are Quantitative, Not Qualitative

IRENE PEPPERBERG

Irene Pepperberg is a research associate in psychology at Harvard University and the author of *The Alex Studies: Cognitive and Communicative Abilities of Grey Parrots.*

I believe that the differences between humans and nonhumans are quantitative, not qualitative.

Perhaps this idea is hardly surprising, coming from someone who has spent her scientific career studying the abilities of (supposedly) small-brained nonhumans. The idea is not exactly new. It may be somewhat controversial: Many of my colleagues spend much of their time searching for the defining difference that separates humans and nonhumans (and they may be right to do so); moreover, the current social and political climate challenges evolution on what seems to be a daily basis. But why is the idea dangerous? Because if we take the idea to its logical conclusion, it challenges almost every aspect of our lives, scientific and nonscientific alike.

Scientifically, the idea challenges the views of many researchers who continue to hypothesize about the next human–nonhuman 'great divide.' Interestingly, however, detailed observation and careful experimentation have repeatedly demonstrated that nonhumans often possess abilities once thought to belong only to humans. Humans, for example, are not the only tool-using species, nor the only tool-making species, nor the only species to act cooperatively.

So one has to wonder to what degree nonhumans share other abilities still considered exclusively human. The critical words here are 'to what degree.' Does one count lack of a particular behavior as a defining criterion, or possibly accept the existence of less complex versions of that behavior as evidence for a continuum? Arguably, I may be just blurring the difference between 'qualitative' and 'quantitative;' if so, so be it. Such blurring will not affect the dangerousness of my idea.

My idea is dangerous because it challenges scientists at a more basic level – that of how we perform research. Let me state clearly that I'm not against animal research. I wouldn't be alive today without it, and I work daily with captive animals that, although domestically bred (and provided, by any standard, with a fairly cushy existence), are still essentially wild creatures denied their freedom.

But if we believe in a continuum, then we must at least question our right to perform experiments on our fellow creatures. We need to think about how to limit animal experiments and testing to what is essential and to insist on humane (note the term!) housing and treatment. And we must accept the significant cost in time, effort, and money thereby incurred – increases that will come at the expense of something else in our society.

My idea, taken to its logical conclusion, is dangerous because it should also affect our choices as to the origins of the clothes we wear and the foods we eat. Again, I'm not campaigning against T-bone steaks and leather shoes; I find that I personally cannot remain healthy on a totally vegetarian diet, and sheepskin boots definitely ease the rigors of a Massachusetts winter.

But if we believe in a continuum, we must at least question our right to use fellow creatures for our sustenance. We need to become aware of, for example, the conditions under which creatures destined for the slaughterhouse live their lives, and we need to learn about and ameliorate the conditions in which their lives are ended. And, again, we must accept the costs involved in such decisions.

If we do not believe in a distinct boundary between humans and nonhumans, we need to rethink other aspects of our lives. Do we have the right to clear-cut forests in which our fellow creatures live? To pollute the air, soil, and water we share with them, solely for our own benefit? Where do we draw the line? Life may be much simpler if we do firmly draw a line, but is simplicity a valid rationale?

And, in case anyone wonders about my own personal view: I believe that humans are the ultimate generalists – creatures that may lack specific talents or physical adaptations that have been finely honed in other species but whose additional brain power enables them, in an exquisite manner, to (for example) integrate information, improvise with what is present, and alter or adapt to a wide range of environments – but that this additional brain power is (and provides) a quantitative, not qualitative, difference.

Groups of People May Differ Genetically in Their Average Talents and Temperaments

STEVEN PINKER

Stephen Pinker is a psychologist at Harvard University. He is the author of many books, including *The Blank Slate*.

The year 2005 saw several public appearances of what I predict will become the dangerous idea of the next decade: that groups of people may differ genetically in their average talents and temperaments.

In January, Harvard president Larry Summers caused a firestorm when he cited research showing that women and men have nonidentical statistical distributions of cognitive abilities and life priorities.

In March, developmental biologist Armand Leroi published an op-ed piece in the *New York Times* rebutting the conventional wisdom that race does not exist. (The conventional wisdom is coming to be known as Lewontin's fallacy: that because most genes may be found in all human groups, the groups don't differ at all. But patterns of *correlation* among genes do differ between groups, and different clusters of correlated genes correspond well to the major races labeled by common sense.)

In June, the *Times* reported a forthcoming study by physicist Gregory Cochran, anthropologist Jason Hardy, and population geneticist Henry Harpending proposing that Ashkenazi Jews have been biologically selected for high intelligence and that their well-documented genetic diseases are a by-product of this evolutionary history.

In September, political scientist Charles Murray published an article in *Commentary* reiterating his argument from *The Bell Curve* that average racial differences in intelligence are intractable and partly genetic.

Whether or not these hypotheses hold up (the evidence for gender differences is reasonably good, for ethnic and racial differences much less so), they are widely perceived to be dangerous. Summers was subjected to months of vilification, and proponents of ethnic and racial differences in the past have been targets of censorship, violence, and comparisons to Nazis. Large swaths of the intellectual landscape have been reengineered to try to rule these hypotheses out *a priori* (race does not exist, intelligence does not exist, the mind is a blank slate inscribed by parents). The underlying fear that reports of group differences will fuel bigotry is not, of course, groundless.

The intellectual tools to defuse the danger are available. 'Is' does not imply 'ought.' Group differences, when they exist, pertain to the average or variance of a statistical distribution rather than to individual men and women. Political equality is a commitment to universal human rights, and to policies that treat people as individuals rather than as representatives of groups; it is not an empirical claim that all groups are indistinguishable. Yet many commentators, to say nothing of the wider world community, seem unwilling to grasp these points.

Advances in genetics and genomics will soon enable us to test hypotheses about group differences rigorously. Perhaps geneticists will forbear from performing these tests, but we shouldn't count on it. The tests could very well emerge as by-products of research in biomedicine, genealogy, and deep history – research that no one wants to stop.

The human genomic revolution has spawned an enormous amount of commentary about the possible perils of cloning and human genetic enhancement. I suspect that these are red herrings. When people realize that cloning is just forgoing a genetically mixed child for a twin of one parent and is not the resurrection of the soul or a source of replacement organs, no one will want to do it. Likewise, when they realize that most genes have costs as well as benefits (they may raise a child's IQ but also predispose him to genetic disease), 'designer babies' will lose whatever appeal they have. But the prospect of genetic tests of group differences in psychological traits is both more likely and more incendiary, and one that the current intellectual community is ill-equipped to deal with.

The Genetic Basis of Human Behavior

J. CRAIG VENTER

J. Craig Venter is founder and president of the J. Craig Venter Institute and the J. Craig Venter Science Foundation, former president of Celera Genomics, and decoder of the human genome.

With our initial analysis of the sequence of the human genome, particularly with the much smaller than expected number of human genes, the genetic determinists seemed to have clearly suffered a setback. After all, those looking for one gene for each human trait and disease won't be happy with as few as twenty thousand or so genes, when hundreds of thousands were anticipated. Deciphering the genetic basis of human behavior has been a complex and largely unsatisfying endeavor because of the limitations of the existing tools of genetic-trait analysis, particularly complex traits involving multiple genes.

All this will soon undergo a revolutionary transformation. The rate of change of DNA-sequencing technology continues at an exponential pace. We are approaching the time when we will go from having a few human genome sequences to complex databases containing tens to hundreds of thousands of complete genomes, then millions. Within a decade, we will begin rapidly accumulating the complete genetic codes of individuals along with their phenotypic repertoires. By performing multifactorial analysis of the DNA sequence variations, along with the comprehensive phenotypic information gleaned from every branch of human investigatory discipline, we will be able to provide for the first time in history quantitative answers to those questions of what is due to genes and what is due to the environment. This is already taking place in cancer research, where we can measure the differences in genetic mutations inherited from our parents versus those resulting from environmental damage. This good news will help transform the treatment of cancer by allowing us to know which proteins need to be targeted.

However, when these powerful new computers and databases are used to help us analyze who we are as humans, will the public, largely ignorant and afraid of science, be ready for the answers we are likely to get?

For example, we know from experiments on fruit flies that there are genes that control many behaviors, including sexual activity. We sequenced the dog genome a couple of years ago and now an additional breed has had its genome decoded. The canine world offers a unique look into the genetic basis of behavior. The large number of distinct dog breeds originated from the wolf genome by selective breeding, yet each breed retains only subsets of the wolf behavioral spectrum. There is a genetic basis not only in the appearance of the breeds (a thirty-fold difference in weight and a sixfold difference in height) but in behavior: For example, border collies use the power of their stare to herd sheep instead of freezing them in place in order to devour them.

We may attribute behaviors in other mammalian species to genes and genetics, but when it comes to humans we seem to like the notion that we're all created equal, that each child is a 'blank slate.' As we obtain the sequences of more and more mammalian genomes, including more human sequences, together with basic observations and common sense, we will be forced to turn away from these politically correct interpretations, as our new genomic tool sets allow us to sort out nature and nurture. We are at the threshold of a realistic biology of humankind.

It will inevitably be revealed that there are strong genetic components associated with most aspects of our human nature: personality subtypes, language capabilities, mechanical abilities, intelligence, sexual activities and preferences, intuitive thinking, quality of memory, will power, temperament, athletic abilities, and so on. We will find unique manifestations of human activity linked to genetics associated with isolated and/or inbred populations.

The danger rests with what we already know – that we are *not* all created equal. Further danger comes with our ability to quantify and measure the genetic side of the equation before we can fully evaluate the environmental components of human nature – a much more difficult task. The genetic determinists appear to be winning, but we cannot let them forget or ignore the wide range of human potential even with our limiting genetic repertoire.

Marionettes on Genetic Strings

JERRY COYNE

Jerry Coyne is an evolutionary biologist and a professor in the
Department of Ecology and Evolution at the University of Chicago. He is
the author (with H. Allen Orr) of *Speciation*.

For me, one idea that is dangerous and possibly true is an
extreme form of evolutionary psychology – the view that many
behaviors of modern humans were genetically hardwired (or
softwired) in our distant ancestors by natural selection.

The reason I say that this idea *might* be true is that we cannot
be sure of the genetic and evolutionary origin of most human
behaviors. It is difficult or impossible to test many of the con-
jectures of evolutionary psychology. Thus, we can say only that
behaviors such as the sexual predilections of men and women
and the extreme competitiveness of males are consistent with
evolutionary psychology.

But consistency arguments have two problems. First, they are
not hard scientific proof. Are we satisfied that sonnets are literary extensions of the phallus simply because some male poets
might have used them to seduce females? Arguments like this
fail to meet the normal standards of scientific evidence.

Second, as is well known, one can make consistency

arguments for virtually every human behavior. Given the possibilities of kin selection (natural selection for behaviors that do no good for their performers but are advantageous to their kin) and reciprocal altruism, and our ignorance of the environments of our ancestors, there is no trait that cannot be explained by some evolutionary story. Indeed, stories have been concocted to explain the evolution of even such manifestly maladaptive behaviors as homosexuality, priestly celibacy, and extreme forms of altruism (e.g., self-sacrifice during wartime). But surely we cannot consider it scientifically proved that genes for homosexuality are maintained in human populations by kin selection. Not only are we ignorant of the genetic basis of behaviors like homosexuality but we also lack any information about how natural selection acted on such genes.

Nevertheless, much of human behavior appears to conform to Darwinian expectations. Males are promiscuous and females coy. We usually treat our relatives better than we do other people. The problem is where to draw the line between those behaviors that are so obviously adaptive that no one doubts their evolutionary origin (e.g., sleeping and eating), those which are probably but not as obviously adaptive (e.g., human sexual behavior and our fondness for fats and sweets), and those whose adaptive basis is purely speculative (e.g., the origin of art or our love of the outdoors).

Although I have been highly critical of evolutionary psychology, this is not from political motives, nor do I think that the discipline is in principle misguided. Rather, I have been critical because evolutionary psychologists seem unwilling to draw lines between what can be taken as demonstrated and what remains speculative – an attitude that has made the discipline more of a faith than a science. This lack of rigor threatens the reputation of all of evolutionary biology, making it seem as if we spend most of our time dreaming up ingenious stories. If we are truly to understand human nature and use this knowledge constructively, we must distinguish through rigorous research the *probably* true from the *possibly* true.

So, why do I see evolutionary psychology as dangerous? I think it is because I am fearful of seeing myself and my fellow humans as mere marionettes dancing on genetic strings. I would like to think we have immense freedom to better ourselves as individuals and create a just and egalitarian society. Granted, genetics is not destiny, but neither are we completely free of our evolutionary baggage. Might genetics really rein in our ability to change? If so, then some claims of evolutionary psychology give us convenient but dangerous excuses for behaviors that seem unacceptable. It is all too easy, for example, for philandering males to excuse their behavior as evolutionarily justified. Evolutionary psychologists argue that it is possible to overcome our evolutionary heritage. But what if it is not so easy to take this Dawkinsian road and 'rebel against the tyranny of the selfish replicators'?

Francis Crick's Dangerous Idea

V. S. RAMACHANDRAN

V. S. Ramachandran is the director of the Center for Brain and Cognition at the University of California, San Diego and an adjunct professor at the Salk Institute, La Jolla. He is the author of *Phantoms in the Brain: Probing the Mysteries of the Human Mind*.

I am a brain, my dear Watson, and the rest of me is a mere appendage.

– SHERLOCK HOLMES

An idea that would be 'dangerous if true' is what Francis Crick referred to as 'the astonishing hypothesis' – the notion that our conscious experience and sense of self consists entirely of the activity of 100 billion bits of jelly, the neurons that constitute the brain. We take this for granted in these enlightened times – but even so, it never ceases to amaze me. Some scholars have criticized Crick's tongue-in-cheek phrase (the title of his last book) on the ground that the hypothesis he refers to is neither astonishing nor a hypothesis, since we already know it to be true. Yet the far-reaching philosophical, moral, and ethical dilemmas it poses have not been recognized widely enough. It is in many ways the ultimate dangerous idea.

Let's put this in historical perspective.

As Freud once pointed out, the history of ideas in the last few centuries has been punctuated by revolutions – major upheavals of thought that have forever altered our view of ourselves and our place in the cosmos. First, the Copernican system dethroned the earth as the center of the cosmos. Second, the Darwinian revolution introduced the idea that, far from being the climax of 'intelligent design,' we are merely neotonous apes that happen to be slightly cleverer than our cousins. Third, the Freudian view taught that even though you claim to be in charge of your life, your behavior is in fact governed by a cauldron of drives and motives of which you are largely unconscious. And fourth, the discovery of DNA and the genetic code implies, to quote James Watson, that '[t]here are only molecules. Everything else is sociology.'

To this list, we can now add a fifth: the neuroscience revolution and its corollary, pointed out by Crick – the astonishing hypothesis that even our loftiest thoughts and aspirations are mere by-products of neural activity. *We are nothing but a pack of neurons.*

If all this seems dehumanizing, you haven't seen anything yet.

This dangerous idea will lead to a philosophical dilemma that will emerge three hundred to five hundred years from now, when we completely understand the brain. But let's speculate: Imagine that today a neuroscientist can transplant your brain into a vat filled with a culture medium and artificially create patterns of activity that will make you feel as though you are living the lives of, say, Francis Crick, Bill Gates, Hugh Hefner, and Mark Spitz, with a dash of Mohandas Gandhi. You – or rather, your brain – will enjoy, in parallel, many of the positive attributes and experiences of these people. At the same time, the neuroscientist makes sure that your brain retains your original identity – including all the memories of your lifetime, strung together by your sense of self. Bear in mind that you experience only certain key aspects of these other lives (and your own), as a result of the right pattern of activity having been created in your brain. But none of it exists in the outside world. It's a delusion of sorts, though one that can't get you into trouble.

Of course, this is the stuff of science fiction, but in my view the idea hasn't been taken to its logical conclusion, nor have its philosophical implications been clearly spelled out. It is possible that the neuroscientist cannot accurately preserve every last detail of 'you' in your entirety, given the slight changes introduced by the addition of the attributes and experiences of those others. But even if he creates a reasonably good approximation of you, my core argument would still be valid. (After all, you already fluctuate from moment to moment!)

Given a choice, would you choose the vat scenario or be content to remain the 'real' you in the real world you live in now. (Assume, for the sake of argument, that the real you is fairly happy and that the chances of eventually dying – or living eternally – are the same, whether you're in the vat or in the real world.) Ironically, most people I know – even scientists – pick the latter alternative, on the grounds that it is 'real.' Yet there is absolutely no rational justification for this choice, because in a sense you already *are* a brain in a vat – a vat called the cranial vault, nurtured by cerebrospinal fluid and bombarded by photons transmitted via the retina. All I've asked you is 'Which vat do you want?' – and you have picked the crummy one! (The most original answer came from my colleague Stuart Anstis, who said, 'The neuroscientist can leave out Gates, Crick, Spitz, and Gandhi; just Hugh Hefner will do.')

There's a sense in which my question poses the ultimate philosophical dilemma. If the logical argument is correct, then the time may come when the world will consist of warehouses full of rows and rows of vats of brains that can be kept alive indefinitely, replete with delightful experiences.

It seems inconceivable that your consciousness and personal memories depend on the actual atoms that now constitute your brain. Surely, they depend entirely on software – that is, the information content. The atoms, after all, are renewed completely every few months, yet you are still 'you.' So in some ultimate sense, you could ask whether it matters which software continues in which vat or how it is instantiated. What if the neuroscientist created several vats, with several brains identical to yours, and put them in several vats: Which one is you? Would 'you' continue in all of them in parallel? It may well be that our ordinary notions about unity and numerosity – and the corresponding terminology – are hopelessly inadequate in dealing with questions about minds and brains. (Just as our everyday notion of causation breaks down in quantum mechanics.)

This raises an even more enigmatic paradox. From an objective, third-person point of view, there's nothing special about the information in your brain, whether in your cranium or in a vat, but from your internal perspective it's everything. The irony is that our brains create an objective science and then proceed to push out subjective experience of the very selves that gave rise to science in the first place! Isn't something wrong here?

These are brain-boggling conundrums. There is only one argument against the vat scenario that I can think of, but it isn't really a logical argument: Every human being is different. Each of us is a cultured ape, whose unique mind has been fashioned by the contingent nature of our life experiences derived from the real, external world. The universe is a network of causation, of which you are one insignificant node – yet one that would be hard to replicate. And even if this could be done, would you want it done? What's so sacred about 'real' reality? This is a question that belongs to the realm of philosophy rather than of science. Science can provide data relevant to the vat question, but not its ultimate answer.

I confess that I, too, would pick the 'real' me, perhaps because of a foolish sentimental attachment to my present reality, or perhaps because I believe, unconsciously, that there *is* 'something else' after all – something priceless about the here-and-now of conscious experience that we simply don't understand. As the bard might have said, if the question had been addressed to him: 'To be me, or not to be me: that is the question.'

Being Alone in the Universe

RODNEY BROOKS

Rodney Brooks is the director of the MIT Computer Science and Artificial Intelligence Laboratory and the author of *Flesh and Machines: How Robots Will Change Us.*

The thing I worry about most, that may or may not be true, is that perhaps the spontaneous transformation from nonliving to living matter is extraordinarily unlikely. We know that it has happened once, but what if we gain lots of evidence over the next few decades that it happens very rarely?

In my lifetime, we can expect to examine the surface of Mars and the moons of the gas giants in some detail. We can also expect to image extrasolar planets within a few tens of light-years to resolutions where we will be able to detect evidence of large-scale biological activity.

What if none of these observations indicates any life whatsoever? What does that do to our scientific belief that life arose spontaneously? It should not change it, but it will make it harder to defend against nonscientific attacks. And wouldn't we be immensely saddened if we were to discover that there is a vanishingly small probability that life will arise even once in any given galaxy?

Being alone in this solar system will not be such a shock, but to be alone in the galaxy – or worse, alone in the universe – would, I think, drive us to despair and back toward religion as our salve.

Life As an Agent of Energy Dispersal

SCOTT D. SAMPSON

Scott D. Sampson is a paleontologist, chief curator of the Utah Museum of Natural History, and an associate professor in the Department of Geology and Geophysics at the University of Utah.

The truly dangerous ideas in science tend to be those that threaten the collective ego of humanity and knock us farther away from a central position within nature. The Copernican Revolution abruptly dislodged humans from the center of the universe. The Darwinian Revolution yanked *Homo sapiens* from the pinnacle of life. Today another menacing revolution sits at the horizon of knowledge, patiently awaiting broad realization by the same egotistical species.

The dangerous idea is this: The purpose of life is to disperse energy.

Many of us are at least somewhat familiar with the second law of thermodynamics, the unwavering propensity of energy to disperse and, in doing so, transition from high-quality to low-quality forms. More generally, as stated by ecologist Eric Schneider, 'nature abhors a gradient,' where a gradient is simply a difference over a distance – for example, in temperature or pressure. Open physical systems – including those of the atmosphere, hydrosphere, and geosphere – all embody this law, being driven by the dispersal of energy, particularly the flow of heat, as they continually attempt to achieve equilibrium. Phenomena as diverse as lithospheric plate motions, the northward flow of the Gulf Stream, and deadly hurricanes are all examples of second-law manifestations.

There is growing evidence that life, the biosphere, is no different. It has often been said that life's complexity contravenes the second law, indicating the work either of a deity or some unknown natural process, depending on one's bias. Yet the evolution of life and the dynamics of ecosystems obey the second law's mandate, functioning in large part to dissipate energy. They do so not by burning brightly and disappearing, like a fire torching a forest, but through stable metabolic cycles that store chemical energy and continuously reduce the solar gradient. Photosynthetic plants, bacteria, and algae capture energy from the sun and form the core of all food webs. Other kinds of lifeforms consume these 'producers,' making the most of the available energy pool.

In a very real sense, then, virtually all organisms, including humans, are sunlight transmogrified, temporary waypoints in the flow of energy. Viewed from a thermodynamic perspective, ecological succession – that is, changes in the species structure of an ecological community over time – is a process that maximizes the capture and degradation of energy. Similarly, the tendency for life to become more complex over the past 3.5 billion years – as indicated by increasing complexity in anatomical forms, metabolic pathways, and trophic interactions, as well as increasing biomass and biodiversity – is not due simply to natural selection, as most evolutionists still argue, but also to nature's efforts to grab more and more of the sun's flow. The slow burn that characterizes life enables ecological systems to persist over deep time, changing in response to external and internal perturbations.

Ecology has been summarized by the pithy statement: 'Energy flows, matter cycles.' Yet this maxim applies equally to complex systems in the nonliving world; indeed, it unites the biosphere with the physical realm. Complex, cycling, swirling systems of matter have a strong tendency to emerge in the face of energy gradients. This recurrent phenomenon may even have been the driving force behind life's origins.

This radical idea is not new, and certainly not mine. Erwin Schrödinger was one of the first to highlight the modern energetic view, as part of his famous 'What Is Life?' lectures in Dublin in 1943. More recently, Jeffrey Wicken, Harold Morowitz, Eric Schneider, and others have taken these concepts considerably further, buoyed by results from a range of studies, particularly within ecology. Schneider and Dorion Sagan provide an excellent summary of this hypothesis in their 2005 book, *Into the Cool.*

The concept of life as energy flow is profound. Just as Darwin fundamentally connected humans to the nonhuman world, a thermodynamic perspective connects life inextricably to the nonliving world. This dangerous idea, once it has been broadly distributed and understood, is likely to provoke reaction from many sectors, including religion and science. The wondrous diversity and complexity of life through time, far from being the product of intelligent design, is a natural phenomenon intimately linked to energy flow within the physical realm.

Moreover, evolution is not driven by the machinations of selfish genes propagating themselves through the millennia.

Rather, ecology and evolution together operate as a highly successful, extremely persistent means of reducing the gradient generated by our nearest star. In my view, evolutionary theory (the process, not the fact of life's common ancestry!) and biology generally are headed for a major overhaul once investigators fully comprehend the notion that the complex systems of earth, air, water, and life are not only interconnected but interdependent, cycling matter in order to maintain the flow of energy.

Although this statement is reductionist and materialist in the sense that it accounts for a broad diversity of phenomena with a single physical process, it must be noted that the idea is entirely mute with regard to spiritual meaning. That is, the word 'purpose,' as applied here, refers solely to naturalistic function – the workings of natural systems. Thus in no way does it exclude other, 'higher' purposes. Nonetheless, the notion of life as an agent of energy flow is likely to have deep effects well outside the boundaries of science. In particular, broad understanding of life's role in dispersing energy has great potential to help humans reconnect to nature at a pivotal moment in our species' history.

We Are Entirely Alone

KEITH DEVLIN

Keith Devlin is a mathematician and the executive director of the Center for the Study of Language and Information at Stanford University. He is the author, most recently, of *The Math Instinct: Why You're a Mathematical Genius (Along with Lobsters, Birds, Cats, and Dogs)*.

Living creatures capable of reflecting on their own existence are a freak accident, existing for one brief moment in the history of the universe. There may be life elsewhere in the universe, but it does not have self-reflective consciousness. There is no God, no Intelligent Designer, no higher purpose to our lives.

Personally, I have never found this possibility particularly troubling, but my experience has been that most people go to considerable lengths to convince themselves that it is otherwise.

Many people find the suggestion dangerous, because they see it as leading to a life devoid of meaning or moral values. They see it as a suggestion full of despair, an idea that makes our lives pointless. I believe that the opposite is the case. As the product of that unique freak accident, finding ourselves able to reflect on and enjoy our conscious existence, the very unlikeliness and uniqueness of our situation surely should make us highly appreciative of what we have.

Life is not just important to us, it is literally everything we have. That makes it, in human terms, the most precious thing there is. That not only gives life meaning for us – something to be respected and revered – but a strong moral code surely follows automatically.

The fact that our existence has no purpose outside that existence is completely irrelevant to the way we live our lives, since we are *inside* our existence. The fact that our existence has no purpose *for the universe* – whatever that means – in no way means that it has no purpose *for us*. We must ask and answer questions about ourselves *within the framework of our existence as what we are.*

Science May Be Running Out of Control

MARTIN REES

Martin Rees is president of the Royal Society and a professor of cosmology and astrophysics and master of Trinity College, Cambridge. He is the author of, among many other books, *Our Final Century: The 50/50 Threat to Humanity's Survival*.

Public opinion surveys (at least in the United Kingdom), while revealing a generally positive attitude toward science, also suggest a widespread worry that it may be 'running out of control.' This idea is a dangerous one because it could be self-fulfilling.

In the twenty-first century, technology will change the world faster than ever – the global environment, our lifestyles, even human nature itself. We are far more empowered by science than any previous generation was. Science offers immense potential, especially for the developing world, but there could be catastrophic downsides. We are living in the first century in which the greatest risks will come from human actions rather than from nature.

Almost any scientific discovery has a potential for evil as well as for good; its applications can be channeled either way, depending on our personal and political choices. We can't accept the benefits without also confronting the risks. The decisions we make, individually and collectively, will determine whether the outcomes of twenty-first-century sciences are benign or devastating.

But there's a real danger that rather than campaigning energetically for optimum policies, we will be lulled into inaction by fatalism – by a belief that science is advancing so fast and is so strongly influenced by commercial and political pressures that nothing we do will make any difference.

The present sharing of resources and effort among the various sciences results from a complicated tension between many extraneous factors, and the balance is suboptimal. This seems so whether we judge in purely intellectual terms or take account of likely benefit to human welfare. Some research has had the inside track and gained disproportionate resources; others, such as studies of the environment, renewable energy sources, biodiversity, and the like, deserve more effort. Within medical research, for example, the focus is disproportionately on cancer and cardiovascular studies, ailments that loom largest in prosperous countries, rather than on the infectious diseases endemic in the tropics.

Choices on how science is applied should be the outcome of debate extending way beyond the scientific community. Far more research and development can be done than we actually want or can afford to do, and there are many applications of science that we should deliberately eschew.

Even if all the world's scientific academies agreed that a specific type of research had a particularly disquieting net downside, and all countries in unison imposed a ban, what are the chances that it could be effectively enforced? In view of the failure to control drug smuggling or homicides, it is unrealistic to expect that when the genie is out of the bottle we can ever be fully secure against the misuse of science. And in our ever more interconnected world, commercial pressures are harder and harder to regulate. The challenges and difficulties of 'controlling' science in this century will be daunting.

Cynics would go further and say that anything that is scientifically and technically possible will be done — somewhere, sometime — despite ethical and prudential objections and whatever the regulatory regime. Whether this idea is true or false, it's an exceedingly dangerous one, because it engenders a despairing pessimism and demotivates efforts to secure a safer and fairer world. The future will best be safeguarded — and science has the best chance of being applied optimally — through the efforts of people who are not fatalistic.

Why I Hope the Standard Model Is Wrong About Why There Is More Matter Than Antimatter

FRANK J. TIPLER

Frank J. Tipler is a professor of mathematical physics at Tulane University, author of *The Physics of Immortality*, and coauthor (with John Barrow) of *The Anthropic Cosmological Principle*.

The standard model of particle physics – a theory of all forces and particles except gravity, and a theory that has survived all tests over the past thirty years – says it is possible to convert matter entirely into energy. Old-fashioned nuclear physics allows some matter to be converted into energy, but because nuclear physics requires the number of heavy particles (like neutrons and protons) and light particles (like electrons) to be separately conserved in nuclear reactions, only a small fraction (less than 1 percent) of the mass of the uranium or plutonium in an atomic bomb can be converted into energy. The standard model says that there is a way to convert all the mass of ordinary matter into energy; for example, it is in principle possible to convert the proton and electron making up a hydrogen atom entirely into energy. Particle physicists have long known about this possibility but have considered it forever irrelevant to human technology, because the energy required to convert matter into pure energy via this process is at the very limit of our most powerful accelerators (a trillion electron volts, or one TeV).

I am very much afraid that the particle physicists are wrong about this standard-model pure-energy conversion process being forever irrelevant to human affairs. I have recently come to believe that the consistency of quantum field theory requires that it should be possible to convert up to 100 kilograms of ordinary matter into pure energy via this process using a device that could fit inside the trunk of a car, a device that could be manufactured in a small factory. Such a device would solve all our energy problems – we would not need fossil fuels – but 100 kilograms of energy is the energy released by a 1,000 megaton nuclear bomb. If such a bomb can be manufactured in a small factory, then terrorists everywhere will eventually have such weapons. I fear for the human race if this comes to pass. I very much hope I am wrong about the technological feasibility of such a bomb.

The Idea That We Understand Plutonium

JEREMY BERNSTEIN

Jeremy Bernstein is a physicist and science writer. He is the author of *Hitler's Uranium Club*, among other books.

The most dangerous idea I have come across recently is the idea that we understand plutonium. Plutonium is the most complex element in the periodic table. It has six different crystal phases between room temperature and its melting point. It can catch fire spontaneously in the presence of water vapor, and if you inhale minuscule amounts you will die of lung cancer. It is the principal element in the 'pits' that are the explosive cores of nuclear weapons. In these pits it is alloyed with gallium. No one knows why this works and no one can be sure how stable this alloy is. These pits, in the thousands, are now decades old. What is dangerous is the idea that they have retained their integrity and can be safely stored into the indefinite future.

The Idea That We Should All Share Our Most Dangerous Ideas

W. DANIEL HILLIS

W. Daniel Hillis is a physicist and computer scientist, the chairman of Applied Minds, Inc., and the author of *The Pattern on the Stone: The Simple Ideas That Make Computers Work*.

I don't share my most dangerous ideas. Ideas are the most powerful forces we can unleash on the world, and they should not be let loose without careful consideration of their consequences. Some ideas are dangerous because they are false, like an idea that one race of humans is more worthy than another, or that one religion has a monopoly on the truth. False ideas like these spread like wildfire and have caused immeasurable harm. They still do. Such false ideas should obviously not be encouraged, but there are also plenty of true ideas that should not be spread – ideas about how to cause terror and pain and chaos, ideas of how to better convince people of things that are not true.

I have often seen otherwise thoughtful people so caught up in such an idea that they seem unable to resist sharing it. To me, the idea that we should all share our most dangerous ideas is itself a very dangerous idea. I hope it never catches on.

The Idea That Ideas Can Be Dangerous

DANIEL GILBERT

Daniel Gilbert is the Harvard College Professor of Psychology at Harvard University and the author of *Stumbling on Happiness*.

'Dangerous' does not mean exciting or bold, it means likely to cause great harm. The most dangerous idea is the only dangerous idea: The idea that ideas can be dangerous.

We live in a world in which people are beheaded, imprisoned, demoted, and censured simply because they have opened their mouths, flapped their lips, and vibrated some air. Yes, those vibrations can make us feel sad or stupid or alienated. Too bad. That's the price of admission to the marketplace of ideas. Hateful, blasphemous, prejudiced, vulgar, rude, or ignorant remarks are the music of a free society, and the relentless patter of idiots is how we know we're in one. When all the words in our public conversation are fair, good, and true, it's time to make a run for the fence.

The Fight Against Global Warming Is Lost

PAUL C. W. DAVIES

Paul C. W. Davies is a physicist and cosmologist at Macquarie University in Sydney, Australia, and the author, most recently, of *How to Build a Time Machine*.

Some countries, including the United States and Australia, have been in denial about global warming. They cast doubt on the science that sets alarm bells ringing. Other countries, such as the United Kingdom, are in a panic and want to make drastic cuts in greenhouse emissions. Both stances are irrelevant, because the fight is a hopeless one. In spite of the recent hike in the price of oil, the stuff is still cheap enough to burn. Human nature being what it is, people will go on burning it until it starts running out and simple economics puts the brakes on. Meanwhile the carbon-dioxide levels in the atmosphere will just go on rising. Even if developed countries rein in their profligate use of fossil fuels, the emerging Asian giants of China and India will more than make up the difference. Rich countries, whose wealth derives from decades of cheap energy, can hardly preach restraint to developing nations trying to climb the wealth ladder. And the obvious solution – massive investment in nuclear energy – has been left too late. The main tragedy of

Chernobyl is not the fifty people killed in the disaster but the twenty-year nuclear paralysis it engendered in Western nations. So continued warming looks unstoppable.

Campaigners for cutting greenhouse emissions scare us by proclaiming that a warmer world is a worse world. My dangerous idea is that it probably won't be.

Some bad things will happen. For example, sea level will rise, drowning some heavily populated or fertile coastal areas. But in compensation, Siberia may become the world's breadbasket. Some deserts may expand; others may shrink. Some places will get drier, others wetter. The evidence that the world will be worse off overall is flimsy. What is certainly the case is that we will have to adjust, and adjustment is always painful. Populations will have to move. In two hundred years, some currently densely populated regions may be deserted. But the population movements over the past two hundred have been dramatic, too. I doubt if anything more drastic will be necessary. Once it dawns on people that yes, the world really is warming up, and no, it doesn't imply Armageddon, then the international agreements like the Kyoto protocol will fall apart.

The idea of giving up the global warming struggle is dangerous because it shouldn't have come to this. Humankind has the resources and the technology to cut greenhouse gas emissions. What we lack is the political will. People pay lip service to environmental responsibility, but they are rarely prepared to put their money where their mouth is. Global warming may turn out to be not so bad after all, but many other acts of environmental vandalism are manifestly reckless – the depletion of the ozone layer, the destruction of rain forests, the pollution of the oceans. Giving up on global warming will set an ugly precedent.

Think Outside the Kyoto Box

GREGORY BENFORD

Gregory Benford is a physicist at the University of California at Irvine and a novelist. His latest novel is *Beyond Infinity*.

Few economists expect the Kyoto accords to attain their goals. With compliance coming only slowly and with three big hold-outs – the United States, China, and India – it seems unlikely to make much difference in overall carbon dioxide increases. Yet all the political pressure is on lessening our fossil fuel burning in the face of fast-rising demand. This pits the industrial powers against the legitimate economic aspirations of the developing world – a recipe for conflict.

Those who embrace the reality of global climate change generally insist that there is only one way out of the greenhouse effect: Burn less fossil fuel, or else! Never mind the economic consequences. But the planet itself modulates its atmosphere through several tricks, and we have tended to ignore most of them. The global problem is simple to explain: We capture more heat from the sun than we radiate away. Mostly this is a good thing; otherwise the mean planetary temperature would hover around freezing. But recent human alterations of the atmosphere have resulted in too much of a good thing.

Two methods are getting little attention: sequestering carbon from the air and reflecting sunlight.

Hide the Carbon

Inevitably, we must understand and control the atmosphere, as part of a grand imperative of directing the entire global ecology. There are several schemes to capture carbon dioxide from the air: Promote tree growth, trap carbon dioxide from power plants in exhaust gas domes, or let carbon-rich organic waste fall into the deep oceans. Increasing forestation is a good, though rather limited, step. Capturing carbon dioxide from power plants costs about 30 percent of the plant output, so it's an economic nonstarter. That leaves the third way.

Imagine you are standing in a Kansas field of ripened corn, staring up into a blue summer sky. Imagine the acre around you extending upward, in a transparent air-filled tunnel soaring all the way to space. That long tunnel holds carbon in the form of invisible gas, carbon dioxide – widely implicated in global climate change. But the corn standing as high as an elephant's eye all around you holds four hundred times as much carbon as there is in man-made carbon dioxide – our villain – in the entire column. Yearly, we manage, through agriculture, far more carbon than is causing our greenhouse dilemma.

Take advantage of that. The leftover corncobs and stalks from our fields can be gathered up, floated down the Mississippi, and dropped into the ocean, sequestering its contained carbon. Below about a kilometer depth, beneath a layer called the thermocline, nothing gets mixed back into the air for a thousand years or more. It's not a permanent solution, but it would buy us and our descendants time to find such answers. And it is inexpensive; cost matters.

The United States has large crop residues. It has also ignored the Kyoto accords, saying that such measures would cost too much. And so they would, if we relied purely on traditional methods, policing energy use and carbon dioxide emissions. Clinton era estimates of such costs were around $100 billion a year, a politically unacceptable sum that led Congress to reject the very notion by a unanimous vote.

But if the United States simply used its farm waste to 'hide' carbon dioxide from our air, complying with Kyoto's standard would cost about $10 billion a year, with no change whatsoever in energy use. The whole planet could do the same. Sequestering crop leftovers could offset about a third of the carbon we put into our air. The carbon dioxide we add to our air will end up in the oceans, anyway, from natural absorption, but not nearly quickly enough to help us.

Reflect Away Sunlight

The planet has maintained its perhaps precarious equilibrium throughout billions of years by editing sunlight with cloud cover. As the oceans warm, water evaporates, forming clouds. These reflect sunlight, reducing the heat below, but just how much depends on cloud thickness, water droplet size, particulate density – a forest of detail.

If our climate starts to vary too much, we could consider deliberately adjusting cloud cover in selected areas to offset unwanted heating. It is not hard to make clouds. Volcanoes and fossil-fuel burning do it all the time, by adding microscopic particles to the air. Cloud cover is a natural mechanism we can augment, and another area where possibility of major change in environmental thinking beckons.

A 1997 U.S. Department of Energy study for Los Angeles showed that planting trees and making blacktop and rooftops lighter colored could significantly cool the city in summer. With minimal costs that get repaid within five years, we can reduce summer midday temperatures by several degrees. This would cut air conditioning costs for the residents, simultaneously lowering energy consumption and lessening the urban heat island effect. Incoming rain clouds would not rise as high above the heat blossom of the city, and so would rain on it less. Instead, clouds would continue inland to drop rain on the rest of Southern California, promoting plant growth. These methods are now under way in Los Angeles, a first experiment.

We can combine this with a cloud-forming strategy. Producing clouds over the tropical oceans is the most effective way to cool the planet on a global scale, since the dark oceans absorb the greater part of the sun's heat. This we should explore now, in case sudden climate changes force us to act quickly.

What makes these ideas dangerous?

They are to those environmentalists who find all such steps suspect – smacking of engineering rather than self-discipline. Yet if Kyoto fails to gather momentum, as seems probable, what else can we do? Turn ourselves into ineffectual Mommy-cop states with endless finger-pointing politics? Try to equally regulate both the rich in their SUVs and Chinese peasants who burn coal for warmth?

Our present conventional wisdom might be termed the Puritan solution ('Abstain, sinners!') and is making only slow, small progress. The Kyoto accords call for the industrial nations to reduce their carbon-dioxide emissions to 7 percent below the

1990 level, and globally we are farther from this goal with every year that passes.

These steps are early measures to help us assume our eventual twenty-first-century role as true stewards of the earth, working alongside nature. Recently Billy Graham declared that since the Bible made us stewards of the earth, we have a holy duty to avert climate change. True stewards use the Garden of Eden's own methods.

Our Planet Is Not in Peril

OLIVER MORTON

Oliver Morton is the chief news and features editor of *Nature* and the author of *Mapping Mars* and *Eating the Sun*.

The truth of this dangerous idea is fairly obvious. Environmental crises are a fundamental part of the history of the earth: There have been sudden and dramatic temperature excursions, severe glaciations, vast asteroid and comet impacts. Yet the earth is still here, unscathed.

There have been mass extinctions associated with some of these events, while other mass extinctions may well have been triggered by subtler internal changes to the biosphere. But none of them seem to have done long-term harm. A lot of interesting species died at the end of the Permian period, 250 million years ago, making the early part of the subsequent Triassic perhaps a little duller than it might have been, but there is no evidence that any fundamentally important Earth processes did not eventually recover. I strongly suspect that not a single basic biogeochemical innovation – the sorts of things that underlie photosynthesis and the carbon cycle, the nitrogen cycle, the sulfur cycle, and so on – has been lost in Earth's 4-billion-year lifetime.

Against this background, the current carbon/climate crisis seems pretty small beer. The change in mean global temperatures seems quite unlikely to be much greater than the regular cyclical change between glacial and interglacial climates. Land-use change is immense, but it's not clear how long that will last, and there are rich seedbanks in the soil that will allow restoration. If fossil-fuel use goes unchecked, carbon-dioxide levels may rise as high as they were in the Eocene, some 50 million years ago, and do so at such a rate that they cause a transient spike in ocean acidity. But they will not stay at those high levels, and the Eocene was not such a terrible place.

The earth doesn't need ice caps or permafrost or any particular sea level. Such things come and go and rise and fall as a matter of course. The planet's living systems adapt and flourish, sometimes in a way that provides negative feedback, occasionally with a positive feedback that amplifies the change. A planet that made it through the massive biogeochemical unpleasantness of the late Permian is in little danger from a doubling (or even quintupling) of the very low carbon-dioxide level that preceded the Industrial Revolution, or from the loss of a lot of forests and reefs, or from the demise of half its species, or from the thinning of its ozone layer at high latitudes.

None of this is to say that we, as people, should not worry about global change; we should worry a lot. Climate change may not hurt the planet but it hurts people. In particular, it will hurt people who are too poor to adapt. Significant climate change will alter rainfall patterns and probably patterns of extreme events as well, in ways that could easily threaten the food security of hundreds of millions of people supporting themselves through subsistence agriculture or pastoralism. It will have a huge effect on the lives of the relatively small number of people in places where sea ice is an important part of the environment (and it seems unlikely that anything we do now can change that). In other, more densely populated places, local environmental and biotic change may have similarly sweeping effects.

Secondary to this, the loss of species, both known and unknown, will be experienced by some as a form of damage that goes beyond any deterioration in ecosystem services. Many people will feel themselves and their world diminished by such extinctions – even when those have no practical consequences – despite the fact that they cannot ascribe an objective value to their loss. One does not have to share the values of these people to recognize their sincerity.

All these effects are excellent reasons to act. Yet many people in the various green movements feel compelled to add on the notion that the planet itself is in crisis, or doomed; that all life on Earth is threatened. In a world where that rhetoric is common, pointing out that this eschatological approach to the environment is baseless can be dangerous.

Since the 1970s, the environmental movement has based much of its appeal on personifying the planet and making it seem like a single entity, then seeking to place it in some ways 'in our care.' It is a very powerful notion and one that benefits from the hugely influential iconographic backing of the first pictures of Earth from space. It has inspired much of the good that the environmental movement has done. The idea that the planet is not in peril could thus undermine the movement's power. This is one reason that people react against the idea so strongly.

If the belief that the planet is in peril were merely wrong, there might be an excuse for ignoring it. But the planet-in-peril idea is an easy target for those who, for various reasons, argue against any action at all on the carbon/climate crisis. Here, bad science is a hostage to fortune. What's worse, the idea distorts environmental reasoning. Emphasizing the nonissue of the health of the planet rather than the real issues of effects that harm people leads to a general preference for averting change rather than adapting to it – even though providing the wherewithal for adaptation will often be the most rational response.

Some environmentalists, and perhaps some environmental reporters, will argue that the inflated rhetoric that trades on the mistaken idea of a planet in peril is necessary in order to keep the show on the road. The idea that people can be more easily persuaded to save the planet (which is not in danger) than their fellow human beings (who are) is an unpleasant and cynical one – another dangerous idea, not least because it may hold some truth. But if putting the planet at the center of the debate is a way of involving everyone, of making us feel that we're all in this together, then one can't help noticing that the ploy isn't working all that well. In the rich nations, many people may indeed believe that the planet is in danger, but they don't believe *they* are in danger, and perhaps as a result they're not clamoring for change loud enough, or in the right way, to bring it about.

There is also a problem of learned helplessness. I suspect people are flattered, in a rather perverse way, by the idea that their lifestyle threatens the whole planet rather than just the livelihoods of millions of people they have never met. But the same sense of scale that flatters may also enfeeble. They may come to think that the problems are too great for them to do anything about.

Rolling carbon/climate issues into the great moral imperative of improving the lives of the poor seems more likely to be a sustainable long-term strategy. The most important thing about environmental change is that it hurts people; the basis of our response should be human solidarity.

The planet will take care of itself.

The Effect of Art Can't Be Controlled or Anticipated

APRIL GORNIK

April Gornik is an artist in New York City (Danese Gallery).

Great art is vulnerable to interpretation, which is one reason it remains stimulating and fascinating for generations. The problem inherent in this is that art can inspire malevolent behavior, as per the notion popularly expressed by Anthony Burgess's (and Stanley Kubrick's) *A Clockwork Orange*. When I was young and aspiring to be a conceptual artist, it disturbed me greatly that I couldn't control the interpretation of my work. When I began painting, it was worse; even I wasn't completely sure of what my art meant. That seemed dangerous for me, personally, at that time. I gradually came not only to respect the complexity and inscrutability of painting and art but to see how it empowers the object. I believe that works of art are animated by their creators and remain able to generate thoughts, feelings, responses. However, the fact is that the exact effect of art can't be controlled or fully anticipated.

A 'Grand Narrative'

DENIS DUTTON

Denis Dutton is a professor of the philosophy of art at the University of Canterbury, New Zealand, and editor of *Philosophy and Literature* and *Arts & Letters Daily*.

The humanities have gone through the rise of Theory in the 1960s, its firm hold on English and literature departments through the 1970s and '80s, followed most recently by its much touted decline and death.

Of course, Theory (capitalization is an English department affectation) never operated as a proper research program in any scientific sense – with hypotheses validated or falsified by experiment or accrued evidence. Theory was a series of intellectual fashion statements, clever slogans, and postures, imported from France in the 1960s, then developed out of Yale and other Theory hot spots. The academic work that Theory spawned was noted more for its chosen jargons, which functioned like secret codes, than for any concern to establish truth or advance knowledge. It was all about careers and prestige.

Truth and knowledge, in fact, were ruled out as quaint illusions. This cleared the way, naturally, for an 'anything goes' atmosphere of academic criticism. In reality, it was anything but anything goes, since the political demands of the period included a long list of stereotyped villains (the West, the Enlightenment, dead white males, even clear writing) to be pitted against mandatory heroines and heros (indigenous peoples, the working class, the oppressed, and so forth).

Though the politics remains as strong as ever in academe, Theory has atrophied, not because it was refuted but because everyone got bored with it. Add to that the absurdly bad writing of academic humanists of the period and episodes like the Sokal Hoax, and the decline was inevitable. Theory academics could with high seriousness ignore rational counterarguments, but for them ridicule and laughter were like water thrown at the Wicked Witch. Theory withered and died.

But wait! Here is exactly where my most dangerous idea comes in. What if it turned out that the academic humanities – art criticism, music and literary history, aesthetic theory, and the philosophy of art – actually had available to them a true, and therefore permanently valuable, theory to organize their speculations and interpretations? What if there really existed a hitherto unrecognized 'grand narrative' that could explain the entire history of creation and experience of the arts worldwide?

Aesthetic experience, as well as the context of artistic creation, is a phenomenon both social and psychological. From the standpoint of inner experience, it can be addressed by evolutionary psychology: the idea that our thinking and values are conditioned by the nearly 2 million years of natural and sexual selection in the Pleistocene.

This Darwinian theory has much to say about the abiding, cross-culturally ascertainable values that human beings find in art. The fascination, for example, that people worldwide feel for the exercise of artistic virtuosity, from Praxiteles to Hokusai to Renée Fleming, is not a social construct but a Pleistocene adaptation (which outside of the arts shows itself in sporting interests everywhere). That calendar landscapes worldwide feature alternating copses of trees and open spaces, often hilly land, water, and paths or riverbanks that wind into an inviting distance is a Pleistocene landscape preference (which shows up in both art history and in the design of public parks everywhere). That soap operas and Greek tragedy all present themes of family breakdown ('She killed him because she loved him') is a reflection of ancient, innate content interests in story telling.

Darwinian theory offers substantial answers to perennial aesthetic questions. It has much to say about the origins of art. It is unlikely that the arts came about at one time or for one purpose; they evolved from overlapping interests based in survival and mate selection in the eighty thousand generations of the Pleistocene. How we scan visually, how we hear, our sense of rhythm, the pleasures of artistic expression and joining with others as an audience, and, not least, how the arts excite us using a repertoire of universal human emotions – all of this and more will be illuminated and explained by a Darwinian aesthetics.

I've encountered stiff academic resistance to the notion that Darwinian theory might greatly improve the understanding of our aesthetic and imaginative lives. There's no reason to worry. The most complete, evolutionarily based explanation of a great work of art, classic or recent, will address its form, its narrative content, its ideology, how it is taken in by the eye or mind – and, indeed, how it can produce a deep, even life-transforming pleasure. But nothing in a valid aesthetic psychology will rob art of its appeal, any more than knowing how we evolved to enjoy fat and sweet makes a piece of cheesecake any less delicious. Nor will a Darwinian aesthetics reduce the complexity of art to simple formulas. It will only give us a better understanding of the greatest human achievements and their effects on us.

In the sense that it would show innumerable careers in the humanities over the last forty years to have been wasted on banal politics and execrable criticism, Darwinian aesthetics is a very dangerous idea indeed. For people who really care about understanding art, it would be a combination of fresh air and strong coffee.

Our Universal Moral Grammar's Immunity to Religion

MARC HAUSER

Marc Hauser is a psychologist and biologist at Harvard University and the author of *Wild Minds: What Animals Really Think*.

Here's an idea based on a few studies, some of which my students and I have conducted: It appears that a wide variety of moral judgments are immune to cultural and demographic variation, including religious background. Controlling for age, people with only a high school education are no different from people with advanced degrees when it comes to judging the permissibility of harming another person in certain contexts. People with religious backgrounds are no different in this regard than atheists and agnostics.

Two further pieces of evidence make these results striking and provide support for the idea that some aspects of our moral psychology are immune to cultural background. First, neither utilitarian nor rule-based/nonconsequentialist perspectives are of help in navigating these dilemmas. For example – to cite the classic trolley problem – a person can flip a switch to prevent a trolley from killing five people by diverting it onto a side track, where it will kill only one. Alternatively, the person can push a man onto the tracks, killing him but saving the five people ahead. The utilitarian option, favored by some religions and cultures, would have subjects always pick saving as many people as possible. The rule-based, or nonconsequentialist, option favored by others would have subjects avoid killing even to forestall the deaths of several other people. But these two theoretical positions fail to resolve the variety of moral dilemmas that people confront, leaving the test subjects in a quandary with respect to delivering logically consistent explanations for their judgments of right and wrong. What looks at first like a rational position, backed by religious and legal doctrine or cultural norms, ends up as inconsistent, irrational blundering. Second, some test subjects are clearly religious whereas others are not, yet their judgments and justifications are in many cases the same. If an atheist or an agnostic provides an incoherent explanation for a particular judgment, so too does a Jew, Catholic, Muslim, or Buddhist.

I think that our evolved moral instincts account for this seemingly universal pattern. Others will argue that it is the insignia of divine creation. I don't think this argument flies. Let's unpack the logic into observations, inferences, and conclusions. For a host of moral dilemmas involving harming or helping others, there are – in the studies conducted so far – no statistically significant differences in the patterns of judgments, regardless of whether or not the test subjects are religious. When people with religious backgrounds judge these cases, their religious doctrine does not provide a set of bullet-proof principles for resolving the dilemmas.

We can interpret this result in two ways: Either a divine power created our universal moral sense or evolution did. At this point, we reach a stalemate, because there is no proof for or against a divine power. But those who think a divine power created our universal moral sense have a problem: How do they explain the observation of universal intuitions regarding harming and helping others and the fact that some religions hold principles that are not universal? If you believe that your religion, with its set of doctrinal principles, is perfectly aligned with a divine power's principles, then you have to agree that the universal incidence of the countervailing intuition is derived from some source other than the divine. Biology would be the logical candidate. Alternatively, you could argue that a divine power *is* the source of the universal moral sense but that religions have simply chosen to live by other principles. But if religions are free to choose in this way, deriving their inspiration from something other than the divine, then much of the motivation and emotion underlying formal religion is in jeopardy. This is an irrational position to uphold.

What is dangerous is not the idea that we are endowed with a moral instinct – a biologically evolved faculty for delivering universal verdicts of right and wrong that is immune to religion and other cultural phenomena. What is dangerous is holding on to an irrational position that starts by equating morality with religion and then moves to an inference that a divine power fuels religious doctrine. This step forces religious people to concede that religious doctrine provides an incoherent account of people's moral judgments.

It's a conclusion that ought to lead people to search for inspiration outside the church. I personally prefer the Darwinian pulpit.

Bertrand Russell's Dangerous Idea

NICHOLAS HUMPHREY

Nicholas Humphrey is School Professor at the Centre for Philosophy of Natural and Social Science at the London School of Economics and the author, most recently, of *Seeing Red: A Study in Consciousness*.

Bertrand Russell's idea, put forward eighty years ago in his *Sceptical Essays*, is about as dangerous as they come. I don't think I can better it: 'I wish to propose for the reader's favourable consideration a doctrine which may, I fear, appear wildly paradoxical and subversive. The doctrine in question is this: that it is undesirable to believe in a proposition when there is no ground whatever for supposing it true.'

Hodgepodge Morality

DAVID PIZARRO

David Pizarro is an assistant professor of psychology at Cornell University.

What some individuals consider a sacrosanct ability to perceive moral truths may instead be a hodgepodge of simpler psychological mechanisms, some of which have evolved for other purposes.

It is increasingly apparent that our moral sense comprises a fairly loose collection of intuitions, rules of thumb, and emotional responses that may have emerged to serve a variety of functions, some of which originally had nothing at all to do with ethics. These mechanisms, when tossed in with our general ability to reason, seem to be how humans come to answer the question of good and evil, right and wrong. Intuitions about action, intentionality, and control, for instance, figure heavily into our perception of what constitutes an immoral act. The emotional reactions of empathy and disgust likewise figure into our judgments of who deserves moral protection and who doesn't. But the ability to perceive intentions probably didn't evolve as a way to determine who deserves moral blame, and the emotion of disgust most likely evolved to keep us safe from rotten meat and feces, not to provide us with information about who deserves moral protection.

Discarding the belief that our moral sense provides a royal road to moral truth is an uncomfortable notion. Most people, after all, are moral realists. They believe that acts are objectively right or wrong, like the solutions to math problems. The dangerous idea is that our intuitions may be poor guides to moral truth and can easily lead us astray in our everyday moral decisions.

We Will Understand the Origin of Life Within the Next Five Years

ROBERT SHAPIRO

Robert Shapiro is professor emeritus and a senior research scientist in the Department of Chemistry, New York University. He is the author of *Planetary Dreams: The Quest to Discover Life Beyond Earth*.

Two very different groups will find this development dangerous, and for different reasons, but the outcome is best explained at the end of my discussion.

Just over a half century ago, in the spring of 1953, a famous experiment brought enthusiasm and renewed interest to the origin-of-life field. Stanley Miller, a graduate student at the University of Chicago mentored by Harold Urey, demonstrated that a mixture of small organic molecules (monomers) could readily be prepared by exposing a mixture of simple gases to an electrical spark. Similar compounds were found in meteorites, which suggested that organic monomers may be widely distributed in the universe. If the ingredients of life could be made so easily, then why could they not just as easily assort themselves to form cells?

That same spring, however, another famous paper was published, by James Watson and Francis Crick. They demonstrated that the heredity of living organisms was stored in a very large molecule called DNA. DNA is a polymer, a substance made by stringing many smaller units together as links are joined to form a long chain.

The clear connection between the structure of DNA and its biological function, and the geometrical beauty of the DNA double helix, led many scientists to consider it the essence of life itself. One flaw, however, spoiled this picture: DNA could store information, but it could not reproduce itself without the assistance of proteins, a different type of polymer. Proteins are also adept at catalyzing many other chemical reactions considered necessary for life. The origin-of-life field became mired in the chicken-or-the-egg question. Which came first: DNA or proteins? An apparent answer emerged when it was found that another polymer, RNA (a cousin of DNA), could manage both heredity and catalysis. In 1986, Walter Gilbert proposed that life began in an RNA world – that is, when an RNA molecule that could copy itself was formed, by chance, in a pool of its own building blocks.

Unfortunately, a half century of chemical experiments have demonstrated that nature has no inclination to prepare RNA, or even the building blocks (nucleotides) that must be linked together to form it. Nucleotides are not formed in Miller-type spark discharges, nor are they found in meteorites. Skilled chemists have prepared nucleotides in well-equipped laboratories and linked them to form RNA, but neither chemists nor laboratories were present when life began on the early earth. The Watson-Crick discovery sparked a revolution in molecular biology but it left the origin-of-life question at an impasse.

Fortunately, an alternative solution to this dilemma has gradually emerged: Neither DNA nor RNA nor protein were necessary for the origin of life. Large molecules dominate the processes of life today, but they were not needed to get it started. Monomers themselves have the ability to support heredity and catalysis. The key requirement is that a suitable energy source be available to assist them in the processes of self-organization. A demonstration of the principle involved in the origin of life would require only that a suitable monomer mixture be exposed to an appropriate energy source in a simple apparatus. We could then observe the very first steps in evolution.

Some mixtures will work, but many others will fail, for technical reasons. Some dedicated effort will be needed in the laboratory to prove this point. Why have I specified five years for this discovery? The unproductive polymer-based paradigm is far from dead, and continues to consume the efforts of the majority of workers in the field. A few years will be needed to entice some of them to explore the other solution. I estimate that several years more (the time for a PhD thesis) might be required to identify a suitable monomer-energy combination and perform a convincing demonstration.

Who would be disturbed if such efforts should succeed? Many scientists have been attracted by the RNA world theory because of its elegance and simplicity. Some of them have devoted decades of their careers in efforts to prove it. They would not be pleased if the physicist Freeman Dyson's description proved to be correct: 'Life began with little bags, the precursors of cells, enclosing small volumes of dirty water containing miscellaneous garbage.'

A very different group would find this development as dangerous as the theory of evolution. Those who advocate creationism and intelligent design would feel that another pillar of their belief system was under attack. They have understood the flaws in the RNA world theory and used them to support their supernatural explanation for life's origin. A successful scientific theory in this area would leave one less task for God to accomplish. The origin of life would be a natural (and perhaps frequent) result of the physical laws that govern the universe. This latter thought falls directly in line with the idea of cosmic evolution, which asserts that events since the Big Bang have moved almost inevitably in the direction of life. No miracle or immense stroke of luck was needed to get it started. If this turns out to be the case, then we should expect to be successful when we search for life beyond this planet. We are not the only life that inhabits this universe.

Understanding Molecular Biology Without Discovering the Origins of Life

GEORGE DYSON

George Dyson is a science historian and the author of *Project Orion* and *Darwin Among the Machines*.

I predict that we will reach a complete understanding of molecular biology and molecular evolution without ever discovering the origins of life.

This suggests either a mystery that science cannot explain, or confirmation that life is merely the collective result of a long series of incremental steps, making it impossible to draw a precise distinction between living and nonliving things.

'The only thing of which I am sure,' argued Samuel Butler in 1880, 'is that the distinction between the organic and inorganic is arbitrary; that it is more coherent with our other ideas, and therefore more acceptable, to start with every molecule as a living thing, and then deduce death as the breaking up of an association or corporation, than to start with inanimate molecules and smuggle life into them.'

Every molecule a living thing? That sounds dangerous to me! But where else can you draw the line?

The Problem with Super Mirrors

MARCO IACOBONI

Marco Iacoboni is a neuroscientist and the director of the Transcranial Magnetic Stimulation Laboratory at the David Geffen School of Medicine at UCLA.

Media violence induces imitative violence. If true, this idea is dangerous for at least two reasons. First, because its implications relate to the issue of freedom of speech; second, because it suggests that our rational autonomy is much more limited than we like to think.

The idea is especially dangerous now, because we have discovered a plausible neural mechanism that can explain why observing violence induces imitation. Moreover, the properties of this neural mechanism – the human mirror neuron system – suggest that imitative violence may not always be a consciously mediated process. The argument for protecting even harmful speech ('speech' in the broad sense, including movies and video games) has typically been that the effects of speech are always subject to the mental intermediation of the listener/viewer. If there is a plausible neurobiological mechanism that suggests that such intermediate steps can be by-passed, this argument is no longer valid.

For more than fifty years, behavioral data have suggested that media violence induces violent behavior in the observers. Meta-data show that the effect-size of media violence is much larger than the effect-size of calcium intake on bone mass or of asbestos exposure to cancer. Still, the behavioral data have been criticized. How is that possible?

Two main types of data have been invoked: controlled laboratory experiments and correlational studies assessing types of media consumed and subsequent violent behavior. The lab data have been criticized as not having enough ecological validity; the correlational data have been criticized as having no explanatory power. As a neuroscientist studying the human mirror neuron system and its relations to imitation, I want to focus on a recent neuroscientific discovery that may explain why the strong imitative tendencies that humans have may lead them to imitative violence when exposed to media violence.

Mirror neurons are cells located in the premotor cortex, the part of the brain relevant to the planning, selection, and execution of actions. In the ventral sector of the premotor cortex, there are cells that fire in relation to specific goal-related motor acts, such as grasping, holding, and bringing to the mouth. Surprisingly, a subset of these cells – what we call mirror neurons – also fire when we observe somebody else performing the same action. The behavior of these cells seems to suggest that the observer is looking at his or her own actions reflected in a mirror while watching somebody else's actions. My group has shown, in several studies, that human mirror neuron areas are also critical to imitation. There is evidence that the activation of this neural system is fairly automatic, thus suggesting that it may bypass conscious mediation. Moreover, mirror neurons also code the intention associated with the observed actions, even though there is not a one-to-one mapping between actions and intentions. (I can grasp a cup because I want to drink or because I want to put it in the dishwasher.) This suggests that the system can indeed code sequences of action (that is, what happens after I grasp the cup), even though only one action in the sequence has been observed.

Some years ago, when we still were a very small group of neuroscientists studying mirror neurons and just starting to investigate the role of mirror neurons in intention understanding, we discussed the possibility of super-mirror neurons. After all, if you have such a powerful neural system in your brain, you also want to have some control or modulatory neural mechanisms. We now have preliminary evidence suggesting that some prefrontal areas have super mirrors.

I think super mirrors come in at least two flavors. One is inhibition of overt mirroring and the other – the one that might explain why we imitate violent behavior, which requires a fairly complex sequence of motor acts – is mirroring of sequences of motor actions. Super-mirror mechanisms may provide a fairly detailed explanation for imitative violence that arises after exposure to media violence.

Cyberdisinhibition

DANIEL GOLEMAN

Daniel Goleman is a psychologist and the author of *Emotional Intelligence.*

The Internet undermines the quality of human interaction, allowing destructive emotional impulses freer rein under specific circumstances. The reason is a neural fluke that results in cyberdisinhibition of brain systems that keep our more unruly urges in check. The tech problem: a major disconnect between the ways our brains are wired to connect and the interface offered in on-line interactions.

Communication via the Internet can mislead the brain's social systems. The key mechanisms are in the prefrontal cortex. These circuits instantaneously monitor you and the other person during a live interaction, automatically guiding your responses so that they are appropriate and smooth and ordinarily inhibiting impulses for actions that would be rude or simply inappropriate – or outright dangerous.

In order for this regulatory mechanism to operate well, you depend on real-time, ongoing feedback from the other person. The Internet has no means of allowing such real-time feedback (other than with rarely used two-way audio/video streams).

That puts our inhibitory circuitry at a loss; there is no signal to monitor from the other person. This results in disinhibition: impulse unleashed.

Such disinhibition seems state specific and typically occurs rarely while people are in positive or neutral emotional states. That's why the Internet works admirably for the vast majority of communication. Rather, this disinhibition becomes far more likely when people feel strong negative emotions. What fails to be inhibited are the impulses those emotions generate.

This phenomenon has been recognized since the earliest days of the Internet – then known as the ARPAnet (for the DoD's Advanced Research Projects Agency) and used chiefly by scientists – as 'flaming': the tendency to send abrasive, angry, or otherwise emotionally 'off' cybermessages. The hallmark of a flame is that the same person would never say the words in the e-mail to the recipient were they face to face. His inhibitory circuits would not allow it – and so the interaction would go more smoothly. Face to face, he might still communicate the same core information, but in a more skillful manner. Off-line and in life, people who flame repeatedly tend to become friendless or get fired (unless they already run the company).

The greatest danger from cyberdisinhibition may be to young people. The prefrontal inhibitory circuitry is among the last parts of the brain to become fully mature, doing so sometime in the twenties. During adolescence there is a developmental lag, with teenagers having fragile inhibitory capacities but fully ripe emotional impulsivity. Strengthening these inhibitory circuits can be seen as the singular task in neural development of the adolescent years.

One way this teenage neural gap manifests on-line is 'cyber bullying,' which has emerged among girls in their early teens.

Cliques of girls post or send cruel, harassing messages to a target girl, who typically is both reduced to tears and socially humiliated. The posts and messages are anonymous, though they become widely known among the target's peers. The anonymity and social distance of the Internet allow an escalation of such petty cruelty to levels rarely found in face-to-face contact; seeing someone cry typically halts bullying among girls, but that inhibitory signal cannot come via the Internet.

A more ominous manifestation of cyberdisinhibition can be seen in the susceptibility of teenagers to being induced to perform sexual acts in front of Web cams for anonymous adult viewers who pay to watch and direct. Apparently hundreds of teenagers have been lured into this corner of child pornography by an equally large audience of pedophiles. The Internet gives strangers access to children in their own homes, and the children are tempted to do things on-line that they would never consider in person.

As with any new technology, the Internet is an experiment in progress. It's time we considered what other such downsides of cyberdisinhibition may be emerging – and time we looked for a technological fix, if possible.

The dangerous thought: the Internet may harbor social perils that our inhibitory circuitry was not evolutionarily designed to handle.

Brains Cannot Become Minds Without Bodies

ALUN ANDERSON

Alun Anderson is a senior consultant at *New Scientist*.

A common image for popular accounts of the mind is a brain in a bell jar. The message is that inside that disembodied lump of neural tissue is everything that is you.

It's a scary image, but misleading. A far more dangerous idea is that brains cannot become minds without bodies, that two-way interactions between mind and body are crucial to thought and health, and that the brain may partly think in terms of the motor actions it encodes for the body's muscles to carry out.

We've probable fallen for disembodied brains because of the academic tendency to worship abstract thought. If we took a more democratic view of the whole brain, we would find far more of it being used for planning and controlling movement than for cogitation. Sports writers get it right when they describe stars of football or baseball as 'geniuses.' Their genius requires massive brain power and a superb body, which is perhaps one better than Einstein.

The 'brain-body' view is dangerous because it requires many scientists to change the way they think: It allows back commonsense interactions between brain and body that medical science feels uncomfortable with, makes more sense of feelings like falling in love, and requires a different approach for people who are trying to create machines with humanlike intelligence. And if all this sounds like mere assertion, there's plenty of interesting research out there to back it up.

Interactions between mind and body come out strongly in the surprising links between status and health. The epidemiologist Michael Marmot's celebrated studies show that the lower you are in the pecking order, the worse your health is likely to be. You can explain away only a small part of the trend by poorer access to health care or poorer food or living conditions. For Marmot, the answer lies in 'the impact over how much control you have over life circumstances.' The important message is that state of mind – perceived status – translates into state of body.

The effect of placebos on health delivers a similar message. Trust and belief are often seen as negative in science and the placebo effect is dismissed as a kind of fraud because it relies on the belief of the patient. But the real wonder is that faith can work. Placebos can stimulate the release of pain-relieving endorphins and affect neuronal firing rates in people with Parkinson's disease.

Body and mind interact too in the most intimate feelings of love and bonding. Those interactions have been best explored in voles where two hormones, oxytocin and vasopressin, are critical. The hormones are released as a result of 'the extended tactile pleasures of mating,' as researchers put it, and hit pleasure centers in the brain that essentially addict sexual partners to each other.

Humans are surely more cerebral. But brain scans of people in love show heightened activity where there are lots of oxytocin and vasopressin receptors. Oxytocin levels rise during orgasm and sexual arousal, as they do from touching and massage. There are defects in oxytocin receptors associated with autism. And the hormone boosts the feeling that you can trust others, which is a key part of intimate relations. In a recent laboratory 'investment game,' many investors would trust all their money to a stranger after a puff of an oxytocin spray.

These few stories show the importance of the interplay of minds and hormonal signals, of brains and bodies. This idea has been taken to a profound level in the studies of the neuroscientist Antonio Damasio, who finds that 'gut feelings' are essential to making decisions. 'We don't separate emotion from cognition like layers in a cake,' says Damasio. 'Emotion is in the loop of reason all the time.'

Indeed, the way in which reasoning is tied to body actions may be counterintuitive. Giacomo Rizzolatti of the University of Parma discovered mirror neurons in a part of the monkey brain responsible for planning movement. These nerve cells fire both when a monkey performs an action (like picking up a peanut) and when the monkey sees someone else do the same thing. Before long, similar systems were found in human brains, too.

The surprising conclusion may be that when we see someone do something, the same parts of our brain are activated, as if we were doing it ourselves. We may know what other people intend and feel by simulating what they are doing within the same motor areas of our own brains. As Rizzolatti puts it, 'The fundamental mechanism that allows us a direct grasp of the mind of others is not conceptual reasoning but direct simulation of the observed events through the mirror mechanism.' Direct grasp of others' minds is a special ability that paves the way for our unique powers of imitation, which in turn have allowed culture to develop.

If bodies and their interaction with brain and planning for action in the world are so central to human kinds of mind, where does that leave the chances of creating an intelligent disembodied mind inside a computer? Perhaps the Turing test will be harder than we think. We may build computers that understand language but cannot say anything meaningful, at least until we can give them 'extended tactile experiences.' To put it another way, computers may not be able to make sense until they can have sex.

What Are People Well Informed *About* in the Information Age?

DAVID GELERNTER

David Gelernter is a computer scientist at Yale University and the author of *The Muse in the Machine.*

Let's date the information age to 1982, when the Internet went into operation and the PC had just been born. What if people have been growing less well informed ever since? What if people have been growing steadily more ignorant ever since the so-called information age began?

Suppose an average U.S. voter, college teacher, fifth-grade teacher, fifth-grade student are each less well informed today than they were in 1995 and less well informed then than in 1985. Suppose, for that matter, they were less well informed in 1985 than in 1965.

If this is indeed the 'information age,' what exactly are people well informed *about*? Video games? Clearly history, literature, philosophy, and scholarship in general are not our specialities. This is some sort of technology age: Are people better informed about science? Not that I can tell. In previous technology ages, there was interest across the population in the era's leading technology.

In the 1960s, for example, all sorts of people were interested in the space program and rocket technology. Lots of people learned a little about the basics – what a service module or translunar injection was, why a Redstone-Mercury vehicle was different from an Atlas-Mercury. All sorts of grade school students, lawyers, housewives, English profs were up on these topics. Today there is *no* comparable interest in computers and the Internet – and no comparable knowledge. 'TCP/IP,' 'routers,' 'Ethernet protocol,' 'cache hits' – these are topics of no interest whatsoever outside the technical community. The contrast is striking.

More Anonymity Is Good

KEVIN KELLY

Kevin Kelly is editor-at-large of *Wired* and the author of *Cool Tools*.

More anonymity is good; that's a dangerous idea.

Fancy algorithms and cool technology make true anonymity in mediated environments more possible today than ever before. At the same time, this techno combo makes true anonymity in physical life much harder. For every step that masks us, we move two steps toward totally transparent unmasking. We have caller ID, but also caller ID Block, and then caller ID-only filters. Coming up: biometric monitoring and little place to hide. A world where everything about a person can be found and archived is a world with no privacy, and therefore many technologists are eager to maintain the option of easy anonymity as a refuge for the private.

However, in every system I have seen where anonymity becomes common, the system fails. The recent taint in the honor of Wikipedia stems from the extreme ease with which anonymous declarations can be put into a highly visible public record. Communities infected with anonymity will either collapse or shift the anonymous to pseudoanonymous, as in eBay, where you have a traceable identity behind an invented nickname. Or voting, where you can authenticate an identity without tagging it to a vote.

Anonymity is like a rare-earth metal. These elements are a necessary ingredient in keeping a cell alive, but the amount needed is a mere hard-to-measure trace. In larger doses, these heavy metals are some of the most toxic substances known. They kill. Anonymity is the same. As a trace element in vanishingly small doses, it's good for the system by enabling the occasional whistleblower or persecuted fringe. But if anonymity is present in any significant quantity, it will poison the system.

There's a dangerous idea circulating that the option of anonymity should always be at hand, and that it is a noble antidote to technologies of control. This is like pumping up the levels of heavy metals in your body to make it stronger.

Privacy can be won only by trust, and trust requires persistent identity, if only pseudo-anonymously. In the end, the more trust the better. Like all toxins, anonymity should be kept as close to zero as possible.

A New Golden Age of Medicine

PAUL W. EWALD

Paul W. Ewald is an evolutionary biologist and the director of the Program in Evolutionary Medicine at the University of Louisville. He is the author of *Plague Time*.

My dangerous idea is that we have in hand most of the information we need to facilitate a new golden age of medicine. And what we don't have in hand we can get fairly readily by wise investment in targeted research and intervention. In this golden age, we should be able to prevent most debilitating diseases in developed and undeveloped countries within a relatively short period of time and with much less money than is generally presumed. This is good news. Why is it dangerous?

One array of dangers arises because ideas that challenge the status quo threaten the livelihood of many. When the many are embedded in powerful places, the threat can be stifling, especially when a lot of money and status are at stake; so it is within the arena of medical research and practice. Imagine what would happen if the big diseases – cancers, arteriosclerosis, stroke, diabetes – were largely prevented.

Big pharmas would become small, because the demand for prescription drugs would drop. The prestige of physicians would decrease, because they would no longer be relied on to prolong life. The burgeoning industry of biomedical research would shrink, because governmental and private funding for it would diminish. Also threatened would be scientists whose sense of self-worth is built on the grant dollars they bring in for discovering minuscule parts of big puzzles. Scientists have been beneficiaries of the lack of progress in recent decades, which has caused leaders such as Harold Varmus, the past head of the National Institutes of Health, to declare that what is needed is more basic research. But basic research has not generated many great advancements in the prevention or cure of disease in recent decades.

The major exception is in the realm of infectious disease, where many important advances were generated from tiny slices of funding. The discovery that peptic ulcers are caused by infections that can be cured with antibiotics is one example; another is the discovery that liver cancer can often be prevented by a vaccine against the hepatitis B virus or by screening blood for hepatitis B and C viruses.

The track record of the past few decades shows that these examples are not quirks. They are part of a trend that goes back over a century, to the beginning of the germ theory. And the accumulating evidence supporting infectious causation of the big bad diseases of modern society repeats the pattern that occurred for diseases that have recently been accepted as caused by infection.

The process of acceptance typically occurs over one or more decades and accords with Schopenhauer's generalization about the establishment of truth: It is first ridiculed, then violently opposed, and finally accepted as self-evident. Just a few groups of pathogens seem to be big players: streptococci, *Chlamydia*, some bacteria of the oral cavity, hepatitis viruses, and herpes viruses. If the correlations between these pathogens and the big diseases of wealthy countries does in fact reflect infectious causation, effective vaccines against these pathogens could contribute in a big way to a new golden age of medicine that could rival the first half of the twentieth century.

The transition to this golden age, however, requires two things: a shift in research effort to identifying the pathogens that cause the major diseases, and development of effective interventions against them. The first would be easy to bring about, by restructuring the priorities of NIH (National Institutes of Health); where money goes, so go the researchers. The second requires mechanisms for putting in place programs that cannot be trusted to the free market, for the same kinds of reasons that Adam Smith gave for national defense. The goals of the interventions do not mesh nicely with the profit motive of the free market. Vaccines, for example, are not very profitable.

Pharmas cannot make as much money by selling one vaccine per person to prevent a disease as they can selling a patented drug like Vioxx which will be administered day after day, year after year, to treat symptoms of an illness that is never cured. And though liability issues are important for such symptomatic treatment, the pharmas can argue forcefully that drugs with nasty side effects provide some benefit even to those who suffer from them most, because the drugs are given not to prevent an illness but to ameliorate it. This sort of defense is less convincing when the victim is a child who develops permanent brain damage from a rare complication of a vaccine given to protect it against a chronic illness it might have acquired decades later.

Another aspect of this new golden age will be the ability to distinguish real threats from pseudo-threats. This will allow us to invest in policy and infrastructure to protect people against real threats without squandering resources and destroying livelihoods in efforts to protect against pseudo-threats. Our present predicament on this front is far from this ideal.

Today experts on infectious diseases, and institutions entrusted to protect and improve human health, sound the alarm in response to each novel threat. Recent fears of a devastating pandemic of bird flu is a case in point. Some of the loudest voices offer a simplistic argument, which is that failing to prepare for the worst-case scenarios is irresponsible and dangerous. This criticism has recently been leveled at me and others who question expert proclamations such as those from the World Health Organization and the Centers for Disease Control.

These proclamations informed us that H5N1 bird flu virus poses an imminent threat of an influenza pandemic similar to or even worse than the 1918 pandemic. I have decreased my popularity in such circles by suggesting that the threat of this scenario is essentially nonexistent. In brief, I argue that the 1918 influenza viruses evolved their unique combination of high virulence and high transmissibility in the conditions at the Western Front of World War I. By transporting contagious flu patients into a series of tightly packed groups of susceptible individuals, personnel fostered transmission from people who were completely immobilized by their illness. Such conditions must have favored the predator-like variants of the influenza virus; these variants would have a competitive edge, because they could ruthlessly exploit a person for their own replication and still get transmitted to large numbers of susceptible individuals.

These conditions have not recurred in human populations

since then, and accordingly we have not had any outbreaks of influenza viruses anywhere near as harmful as those that emerged at the Western Front. As long as we do not let such conditions occur again, we have little to fear from a re-evolution of such a predatory virus.

The fear of a 1918-style pandemic fueled preparations by a government that, embarrassed by its failure to deal adequately with the damage from Hurricane Katrina, seems determined to prepare for any perceived threat to save face. I would have no problem with the accusation of irresponsibility if preparations for a 1918-style pandemic were cost-free. But they are not. The $7 billion that the Bush administration sees as a down payment for pandemic preparedness has to come from somewhere. If money is spent to prepare for an imaginary pandemic, our progress could be impeded on other fronts that could lead to, or have already established, real improvements in public health.

Conclusions about the responsibility or irresponsibility of this argument require that the threat from pandemic influenza be assessed relative to the damage that results from the procurement of money from other sources. The only reliable evidence of the damage from pandemic influenza under normal circumstances is the experience of the two pandemics that have occurred since 1918 – one in 1957, the other in 1968. The mortality caused by these pandemics was 1/10 to 1/100 the death toll from the 1918 pandemic.

We do need to be prepared for an influenza pandemic of the normal variety, just as we needed to be prepared for Category 5 hurricanes in the Gulf of Mexico. If possible, our preparations should allow us to stop an incipient pandemic before it materializes. In contrast to many of the most vocal experts, I do not conclude that our surveillance efforts will be quickly overwhelmed by a highly transmissible descendant of H5N1, the influenza virus that has generated the most recent fright. The transition of the H5N1 virus to a pandemic virus would require evolutionary change.

The dialog about this, however, continues to neglect the primary mechanism of the evolutionary change, natural selection. Instead, it is claimed that H5N1 could mutate to become a full-fledged human virus both highly transmissible and highly lethal. Mutation provides only the variation on which natural selection acts. We must consider natural selection if we are to make meaningful assessments of the danger posed by the H5N1 virus.

The evolution of the 1918 virus was gradual; evidence and theory both lead to the conclusion that any evolution of increased transmissibility of H5N1 from human to human will be gradual, as it was with SARS. With surveillance, we can detect such changes in humans and intervene to stop further spread, as was done with SARS. We do not need to trash the economy of Southeast Asia each year to accomplish this.

The dangerous vision of a golden age does not leave the poor countries behind. As I have discussed in my articles and books, we should be able to control much of the damage caused by the major killers in poor countries by infrastructural improvements that not only reduce the frequency of infection but also cause the infectious agents to evolve toward benignity. This integrated approach offers the possibility of remodeling our current efforts against the major killers – AIDS, malaria, tuberculosis, dysentery, and the like. We should be able to move from just holding our ground to instituting the changes that created the freedom from acute infectious diseases enjoyed by inhabitants of rich countries over the past century.

Dangerous indeed! Excellent solutions are often dangerous to the status quo, because they work. One measure of danger to the few but success for the many is the extent to which highly specialized researchers, physicians, and other health care workers will need to retrain, and the extent to which hospitals and pharmaceutical companies will need to downsize. That is what happens when we introduce excellent solutions to health problems. We need not be any more concerned about these difficulties than we were at the loss of the iron lung industry and the retraining of polio therapists and researchers in the wake of the Salk vaccine.

Using Medications to Change Personality

SAMUEL BARONDES

Samuel Barondes is the director of the Center for Neurobiology and Psychiatry at the University of California, San Francisco, and the author of *Better Than Prozac: Creating the Next Generation of Psychiatric Drugs*.

Personality – the pattern of thoughts, feelings, and actions that is typical of each of us – is generally formed by early adulthood. But many people still want to change. Some, for example, consider themselves too gloomy and uptight and want to become more cheerful and flexible. Whatever their aims, they often turn to therapists, self-help books, and religious practices.

In the past few decades, certain psychiatric medications have become an additional tool for those seeking control of their lives. Initially designed to be used for a few months to treat episodic psychological disturbances such as severe depression, they are now being widely prescribed for indefinite use to produce sustained shifts in certain personality traits. Prozac is the best known of these, but many others are on the market or in development. By directly affecting brain circuits that control emotions, these medications can produce desirable effects that may be hard to replicate by sheer force of will or behavioral exercises. Millions take them continually, year after year, to modulate personality.

Nevertheless, the idea of using such drugs to change personality is still dangerous – and not because manipulation of brain chemicals is intrinsically cowardly, immoral, or a threat to the social order. On the contrary, many people feel that they have the opposite effect, helping to increase personal responsibility. The reason for caution is that there have not been any controlled studies of the influence of these drugs on personality over the many years that some people take them. So this is a dangerous idea that can and should be tested, to find out if such sustained drug use is really helpful and if this practice should be continued.

Drugs May Change the Patterns of Human Love

HELEN FISHER

Helen Fisher is a research professor in the Department of Anthropology at Rutgers University and the author of *Why We Love*.

Serotonin-enhancing antidepressants, such as Prozac and many others, can jeopardize feelings of romantic love, feelings of attachment to a spouse or partner, one's fertility, and one's genetic future.

I am working with psychiatrist Andy Thomson on this topic. We base our hypothesis on patient reports, fMRI studies, and other data on the brain. Foremost, as SSRIs (for 'selective serotonin reuptake inhibitors') elevate serotonin, they also suppress dopaminergic pathways in the brain. Because romantic love is associated with elevated activity in dopaminergic pathways, it follows that SSRIs can jeopardize feelings of intense romantic love. SSRIs also curb obsessive thinking – a central characteristic of romantic love – and blunt the emotions. One patient described this reaction well, writing:

> After two bouts of depression in ten years, my therapist recommended I stay on serotonin-enhancing antidepressants indefinitely. As appreciative as I was to have regained

my health, I found that my usual enthusiasm for life was replaced with blandness. My romantic feelings for my wife declined drastically. With the approval of my therapist, I gradually discontinued my medication. My enthusiasm returned and our romance is now as strong as ever. I am prepared to deal with another bout of depression if need be, but in my case the long-term side effects of antidepressants render them off limits.

SSRIs also suppress sexual desire, sexual arousal, and orgasm in as many as 73 percent of users. These sexual responses evolved to enhance courtship and mating. Orgasm produces a flood of oxytocin and vasopressin, chemicals associated with feelings of attachment and pair-bonding behaviors. Orgasm is also a device by which women assess potential mates. Women do not reach orgasm with every coupling, and the 'fickle' female orgasm is now regarded as an adaptive mechanism by which women distinguish males who are willing to expend time and energy to satisfy them. The onset of female anorgasmia may jeopardize the stability of a long-term mateship as well.

Men who take serotonin-enhancing antidepressants also inhibit evolved mechanisms for mate selection, partnership formation, and marital stability. The penis stimulates to give pleasure and advertise the male's psychological and physical fitness; it also deposits seminal fluid in the vaginal canal, fluid that contains dopamine, oxytocin, vasopressin, testosterone, estrogen, and other chemicals that most likely influence a female partner's behavior.

These medications can also influence your genetic future. Serotonin increases prolactin by stimulating prolactin-releasing factors. Prolactin can impair fertility by suppressing the release of hypothalamic GnRH (gonadotropin-releasing hormones) and pituitary FSH (follicle-stimulating hormones) and LH (luteinizing hormones) and/or suppressing ovarian hormone production. Clomipramine, a strong serotonin-enhancing antidepressant, adversely affects sperm volume and motility.

I believe that *Homo sapiens* has evolved (at least) three primary distinct yet overlapping neural systems for reproduction. The sex drive evolved to motivate ancestral men and women to seek sexual union with a range of partners; romantic love evolved to enable them to focus their courtship energy on a preferred mate, thereby conserving mating time and energy; attachment evolved to enable them to rear a child through infancy together. The complex and dynamic interactions between these three brain systems suggest that any medication that changes their chemical checks and balances is likely to alter an individual's courting, mating, and parenting tactics, ultimately affecting fertility and thus genetic future.

The reason my idea is a dangerous one is that the huge drug industry is heavily invested in selling these drugs. Millions of people worldwide currently take them, and as these drugs become generic many more people will take them – inhibiting their ability to fall in love and stay in love. And if patterns of human love subtly change, all sorts of social and political atrocities can escalate.

A Marriage Option for All

DAVID G. MYERS

David G. Myers is a social psychologist at Hope College and the coauthor (with Letha Scanzoni) of *What God Has Joined Together?: A Christian Case for Gay Marriage*.

Much as most scientists have felt compelled by evidence to believe in human evolution or the warming of the planet, I feel compelled by evidence to believe (1) that sexual orientation is a natural, enduring disposition and (2) that the world would be a happier and healthier place if, for all people, romantic love, sex, and marriage were a package.

In my Midwestern social and religious culture, the words 'for all people' transform a conservative platitude into a dangerous idea, over which we are fighting a culture war. On one side are traditionalists, who feel passionately about the need to support and renew traditional marriage. On the other side are progressives, who assume that our sexual orientation is something we did not choose and cannot change, and that we all deserve the option of life within a covenant partnership.

I foresee a bridge across this divide, as folks on both the left and the right engage the growing evidence of our panhuman longing for belonging, of the benefits of marriage, and of the biology and persistence of sexual orientation. We now have lots of data showing that marriage is conducive to healthy adults, thriving children, and flourishing communities. We also have a dozen discoveries of gay-straight differences, in everything from brain physiology to skill at mentally rotating geometric figures. And we have an emerging professional consensus that sexual-reorientation therapies seldom work.

More and more young adults – tomorrow's likely majority, given generational succession – are coming to understand this evidence and to support what in the future will not seem so dangerous: a marriage option for all.

Choosing the Sex of One's Child

DIANE F. HALPERN

Diane F. Halpern is a professor of psychology at Claremont McKenna College and the author of *Sex Differences in Cognitive Abilities and Thought and Knowledge: An Introduction to Critical Thinking.*

For an idea to be truly dangerous it needs to have a strong and near universal appeal. The idea of being able to choose the sex of one's own baby is just such an idea.

Anyone who has a deep-seated and profound preference for a son or daughter knows that this preference may not be rational and that it may represent a prejudice better left unacknowledged. It is easy to dismiss the ability to decide the sex of one's baby as inconsequential. It is already medically feasible for a woman or couple to choose the sex of a baby that has not yet been conceived. There are a variety of safe methods available, such as Preimplanted Genetic Diagnosis (PGD), so named because it was originally designed for couples with fertility problems, not for the purpose of selecting the sex of the child. In PGD, embryos are created in a petri dish, tested for gender, and then implanted into the womb. The pro argument is simple: If the parents-to-be are adults, why not? People have always wanted to be able to choose the sex of their children.

There are ancient records of medicine men and wizened women with various herbs and assorted advice about what to do to have (usually) a son. So what should it matter if modern medicine can finally deliver what old wives' tales have promised for countless generations? Couples won't have to have a 'wasted' child, such as a second child the same sex as the first one, when they really wanted 'one of each.' If a society has too many boys for a while, who cares? The shortage of females will make females more valuable, and the market economy will even things out in time. Meanwhile families will 'balance out,' each one the ideal composition as desired by the adults in the family.

Every year for the last two decades I have asked students in my college classes to write down the number of children they would like to have and the order in which they ideally want to have girls and boys. I have taught in several different countries (Turkey, Russia, Mexico) and types of universities, but despite large differences the modal response is two children, first a boy, then a girl. If students reply that they want one child, it is most often a boy; if it is three children, they are most likely to want a boy, then a girl, then a boy. The students in my classes are not a random sample of the population: They are well educated and more likely to hold egalitarian attitudes than the general population. Yet if they acted on their stated intentions, even they would have an excess of male firstborns and an excess of males overall. In a short time, those personality characteristics associated with being either an only child or firstborn and those associated with being male would be so confounded that it would be difficult to separate them.

The excess of males that would result from allowing every mother or couple to choose the sex of the next baby would *not* correct itself at the societal level, because at the individual level the preference for sons is stronger than the market forces of supply and demand. The evidence for this conclusion comes from many sources (UNICEF among them), including regions of the world where the ratio of young women to men is so low that it could have been caused only by selective abortion and female infanticide. In some regions of rural China, there are so few women that wives are imported from the Philippines and men move to far cities to find women to marry. In response, the Chinese government is now offering a variety of education and cash incentives to families with multiple daughters. There are still few daughters being born in these rural areas, where prejudice against girls is stronger than government incentives and mandates. In India, the number of abortions of female fetuses has increased since sex-selective abortion was made illegal in 1994. The desire for sons is even stronger than the threat of legal action.

In the United States, the data that show preferences for sons are more subtle than the disparate ratios of females and males found in other parts of the world, but the preference for sons is still strong. Because of space limitations, I list only a few of the many indicators that parents in the United States prefer sons: Families with two daughters are more likely to have a third child than families with two sons; unmarried pregnant women who undergo ultrasound to determine the sex of the unborn child are less likely to be married at the time of the child's birth when the child is a girl than when it is a boy; and divorced women with a son are more likely to remarry than divorced women with a daughter.

Perhaps the only ideas more dangerous than that of choosing the sex of one's child would be trying to stop medical science from making advances that allow such choices or allowing the government to control the choices we can make as citizens. There are many important questions to ponder, including how to find creative ways to reduce or avoid negative consequences from even more dangerous alternatives. Consider, for example, what our world would be like if there were substantially more men than women. What if only the rich, or only those who live in rich countries, were able to choose the sex of their children? Is it likely that an approximately equal number of boys and girls would be or could be selected? If not, could a society or should a society make equal numbers of girls and boys a goal?

I am guessing that many readers of child-bearing age want to choose the sex of their as yet unconceived children and reason that there is no harm in this practice. And if you could also choose intelligence, height, hair color, would you add those, too? But then there are few things in life as appealing as the possibility of a perfectly balanced family, which according to the modal response means an older son and younger daughter, looking just like an improved version of you.

The Idea of Ideas

SETH LLOYD

Seth Lloyd is a quantum computer scientist at MIT and the author of *Programming the Universe: A Quantum Computer Scientist Takes on the Cosmos.*

The most dangerous idea is the genetic breakthrough that made people capable of ideas in the first place. The idea of ideas is nice enough in principle, and ideas certainly have had their impact for good. But one of these days, one of those nice ideas is likely to have the unintended consequence of destroying everything we know.

Meanwhile we cannot not stop creating and exploring new ideas: The genie of ingenuity is out of the bottle. To suppress the power of ideas will hasten catastrophe, not avert it. Rather, we must wield that power with the respect it deserves.

Who risks no danger reaps no reward.

The Human Brain Will Never Understand the Universe

KARL SABBAGH

Karl Sabbagh is a writer and television producer and the author of *The Riemann Hypothesis: The Greatest Unsolved Problem in Mathematics.*

Our brains may never be well enough equipped to understand the universe, and we are fooling ourselves if we think they will.

Why should we expect to be able eventually to understand how the universe originated, evolved, and operates? While human brains are complex and capable of many amazing things, there is not necessarily any match between the complexity of the universe and the complexity of our brains, any more than a dog's brain is capable of understanding every detail of the world of cats and bones, or the dynamics of stick trajectories when thrown. Dogs get by and so do we, but do we have a right to expect that the harder we puzzle over these things the nearer we will get to the truth? Recently I stood in front of a three-meter-high model of the Ptolemaic universe in the Museum of the History of Science in Florence, and I remembered how well that worked as a representation of the motions of the planets until Copernicus and Kepler came along.

Nowadays, no element of the theory of giant interlocking cogwheels at work is of any use in understanding the motions of the stars and planets (and indeed Ptolemy himself did not argue that the universe really was run by giant cogwheels). Ockham's razor is used to compare two theories and allow us to choose which is more likely to be 'true,' but hasn't it become a comfort blanket whenever we are faced with aspects of the universe that seem unutterably complex – string theory, for example? But is string theory just the Ptolemaic clockwork *de nos jours*? Can it be succeeded by some simplification, or might the truth be even more complex and far beyond the neural networks of our brain to understand?

The history of science is littered with examples of two types of knowledge advancement. There is imperfect understanding that 'sort of' works and is then modified and replaced by something that works better without destroying the validity of the earlier theory. Newton's theory of gravitation was replaced by Einstein. Then there is imperfect understanding that is replaced by some new idea that owes nothing to older ones. Phlogiston theory, the ether, and so on are replaced by ideas that save the phenomena, lead to predictions, and convince us that they are nearer the truth. Which of these categories really covers today's science? Could we be fooling ourselves by playing around with modern phlogiston?

And even if we are on the right lines in some areas, how much of what there is to be understood in the universe do we really understand? Fifty percent? Five percent? The dangerous idea is that perhaps we understand half a percent and all the brain and computer power we can muster may take us up to one or two percent in the lifetime of the human race.

Paradoxically, we may find that the only justification for

pursuing scientific knowledge is for the practical applications it leads to – a view that runs contrary to the traditional support of knowledge for knowledge's sake. And why is this paradoxical? Because the most important advances in technology have come out of research that was not seeking to develop those advances but to understand the universe.

So if my dangerous idea is right – that the human brain and its products are actually incapable of understanding the truths about the universe – it will not (and should not) lead to any diminution in our attempts to do so. Which means, I suppose, that it's not really a dangerous idea at all.

The World May Be Fundamentally Inexplicable

LAWRENCE M. KRAUSS

Lawrence M. Krauss is Ambrose Swasey Professor of Physics and Astronomy and director of the Center for Education and Research in Cosmology and Astrophysics at Case Western Reserve University and the author, most recently, of *Hiding in the Mirror: The Mysterious Allure of Extra Dimensions, from Plato to String Theory and Beyond.*

Science has progressed for four hundred years by ultimately explaining observed phenomena in terms of fundamental theories that are rigid. Even minor deviations from the predicted behavior are not allowed by a theory; if such deviations are observed, these provide evidence that the theory must be modified, usually being replaced by a yet more comprehensive theory that fixes a wider range of phenomena.

The ultimate goal of physics, as it is often described, is to have a 'theory of everything,' in which all the fundamental laws that describe nature can neatly be written down on the front of a T-shirt (even if the T-shirt can exist only in ten dimensions). However, with the recognition that the dominant energy in the universe resides in empty space – something so peculiar that it appears very difficult to understand within the context of any theoretical ideas we now possess – more physicists have been exploring the idea that perhaps physics is an 'environmental science,' that the laws of physics we observe are mere accidents of our circumstances, and that there could exist an infinite number of different universes with different laws of physics.

This would be true even if there were some candidate for a fundamental mathematical-physical theory. For example, as in an idea currently in vogue related to string theory, perhaps the fundamental theory allows an infinite number of different 'ground state' solutions, each of which describes a different possible universe with a consistent set of physical laws and physical dimensions.

It may be that the only way to understand why the laws of nature we observe in our universe are the way they are is to understand that if they were any different, then life could not have arisen in our universe and we would thus not be here to observe them.

This is one version of the infamous anthropic principle. But things could actually be worse: It's equally likely that many different combinations of laws would allow life to form, and that it's a pure accident that the constants of nature result in the combinations we experience in our universe. Or it could be that the mathematical formalism is so complex that the ground states of the theory – the set of possible states that might describe our universe – might not be determinable.

The end of 'fundamental' theoretical physics (the search for fundamental microphysical laws – there will still be lots of work for physicists who investigate the host of complex phenomena at larger scales) might very well occur not with a theory of everything but with the recognition that all so-called fundamental theories that describe nature are purely phenomenological – that is, derivable from observational phenomena – and don't reflect any underlying grand mathematical structure of the universe which would allow a basic understanding of why the universe is the way it is.

The 'Landscape'

LEONARD SUSSKIND

Leonard Susskind is a theoretical physicist at Stanford University and the author of *The Cosmic Landscape: String Theory and the Illusion of Intelligent Design*.

I have been accused of advocating an extremely dangerous idea.

According to some people, the landscape idea will eventually ensure that the forces of intelligent design (and other unscientific religious ideas) will triumph over true science. From one of my most distinguished colleagues: 'From a political, cultural point of view, it's not that these arguments are religious but that they denude us from our historical strength in opposing religion.'

Others have expressed the fear that my ideas, and those of my friends, will lead to the end of science (methinks they overestimate me). One physicist calls it 'millennial madness.' And from another quarter, Christoph Schönborn, Cardinal Archbishop of Vienna, has accused me of 'an abdication of human intelligence.'

As you may have guessed, the idea in question is the anthropic principle, which seeks to explain the laws of physics and the constants of nature by saying, 'If the laws of nature were different, intelligent life would not exist to ask why the laws of nature are what they are.'

On the face of it, the anthropic principle is far too silly to be dangerous. It sounds no more sensible than explaining the evolution of the eye by saying that if it hadn't evolved, there would be no one to read this page. But the anthropic principle is really shorthand for a rich set of ideas that are beginning to influence and even dominate the thinking of almost all serious theoretical physicists and cosmologists.

Let me strip the idea down to its essentials. Without all the philosophical baggage, what it says is straightforward: The universe is vastly bigger than the portion we can see, and on a very large scale it is as varied as possible. In other words, rather than being a homogeneous, mono-colored blanket, it is a crazy-quilt patchwork of different environments. This is not an idle speculation. There is a growing body of empirical evidence confirming the inflationary theory of cosmology, which underlies the hugeness and hypothetical diversity of the universe.

Meanwhile string theorists, much to the regret of many of them, are discovering that the number of possible environments described by their equations is far beyond millions or billions. This enormous space of possibilities, whose multiplicity may exceed 10500, is called the landscape. If these things prove true, then some features of the laws of physics (maybe most) will be local environmental facts rather than written-in-stone laws – laws that could not be otherwise. The explanation of some numerical coincidences will necessarily be that most of the multiverse is uninhabitable but in some very tiny fraction of it, conditions are fine-tuned enough for intelligent life to form.

That's the dangerous idea and it is spreading like a cancer.

Why is it that so many physicists find these ideas alarming?

Well, they *do* threaten physicists' fondest hope – the hope that some extraordinarily beautiful mathematical principle will be discovered that would completely and uniquely explain every detail of the laws of particle physics (and therefore nuclear, atomic, and chemical physics). The enormous landscape of possibilities inherent in our best theory seems to dash that hope.

What further worries many physicists is that the landscape may be so rich that almost anything can be found – any combination of physical constants, particle masses, and so forth. This, they fear, would eliminate the predictive power of physics. Environmental facts are nothing more than environmental facts. They worry that if everything is possible, there will be no way to falsify the theory – or, more to the point, no way to confirm it. Is the danger real? We shall see.

Another danger that some of my colleagues perceive is that if we 'senior physicists' allow ourselves to be seduced by the anthropic principle, young physicists will give up looking for the 'true' reason for things, the beautiful mathematical principle. My guess is that if the young generation of scientists is really that spineless, then science is doomed anyway. But as we know, the ambition of all young scientists is to make fools of their elders.

And why does Cardinal Archbishop Schönborn find the landscape and the multiverse so dangerous. I will let him explain it himself:

> Now, at the beginning of the twenty-first century, faced with scientific claims like neo-Darwinism and the multiverse hypothesis in cosmology invented to avoid the overwhelming evidence for purpose and design found in modern science, the Catholic Church will again defend

human nature by proclaiming that the immanent design evident in nature is real. Scientific theories that try to explain away the appearance of design as the result of 'chance and necessity' are not scientific at all, but, as John Paul put it, an abdication of human intelligence.

Abdication of human intelligence? No, it's called science.

Seeing Darwin in the Light of Einstein; Seeing Einstein in the Light of Darwin

LEE SMOLIN

Lee Smolin is a theoretical physicist at the Perimeter Institute of Theoretical Physics, in Waterloo, Ontario. He is the author, most recently, of *The Trouble with Physics: The Rise of String Theory, the Fall of a Science, and What Comes Next*.

The revolutions made by Einstein and Darwin are closely related, and their combination will increasingly come to define how we see our worlds – physical, biological, and social.

Before Einstein, the properties of elementary particles were understood as being defined against an absolute, eternally fixed background. This way of doing science had been introduced by Newton. His method was to posit the existence of an absolute and eternal background structure against which the properties of things were defined. Particles have properties defined not with respect to one another but each with respect only to the absolute background of space and time. Einstein's great achievement was to realize successfully the contrary idea, called relationalism, according to which the world is a network of relationships that evolve in time. There is no absolute background, and the properties of something are defined only in terms of its participation in this network of relations.

Before Darwin, species were thought of as eternal categories, defined *a priori*; after Darwin, species were understood to be relational categories – that is, defined in terms of their relationship with the network of interactions making up the biosphere. Darwin's great contribution was to understand that there is a process – natural selection – that can act on relational properties, leading to the birth of genuine novelty by creating networks of relationships that are increasingly structured and complex.

Seeing Darwin in the light of Einstein, we understand that all the properties a species has in modern biology are relational. There is no absolute background in biology.

Seeing Einstein in the light of Darwin suggests that natural selection could act not only on living things but on the properties defining the various species of elementary particles.

At first, physicists thought that the only relational properties an elementary particle might have were its position and motion in space and time. The other properties, like mass and charge, were thought of in the old framework, defined by a background of absolute law. The standard model of particle physics taught us that some of those properties, like mass, are only the consequence of a particle's interactions with other fields. The mass of a particle is determined environmentally, by the phase of the other fields it interacts with.

I don't know which model of quantum gravity is right, but all the leading candidates – string theory, loop quantum gravity, and others – teach us that all properties of elementary particles may be relational and environmental. In different possible universes, there may be different combinations of elementary particles and forces; indeed, all that used to be thought of as fundamental – the elementary particles and space itself – are increasingly seen, in models of quantum gravity, as emergent from an even more elementary network of relations.

The basic method of science after Einstein seems to be to identify something in your theory that is playing the role of an absolute background – that is needed to define the laws that govern objects in your theory – and understand it more deeply as a contingent property which itself evolves subject to law.

For example, before Einstein the geometry of space was thought of as specified absolutely as part of the laws of nature. After Einstein, we understand that geometry is contingent and dynamical, which means it evolves subject to law. Einstein's move can even be applied to aspects of what were thought to be the laws of nature, so that even those aspects turn out to evolve in time.

The basic method of science after Darwin seems to be to identify some property once thought to be absolute and defined *a priori* and recognize that it can be understood because it has evolved by a process of, or a process akin to, natural selection. This has revolutionized biology and is in the process of doing the same to the social sciences.

These two methods are closely related. Einstein emphasizes the relational aspect of all properties described by science, while Darwin proposes that ultimately the law governing the evolution of everything else – including perhaps what were once seen to be laws – is natural selection.

Should Darwin's method be applied even to the laws of physics? Recent developments in elementary-particle physics give us little alternative, if we are to have a rational understanding of the laws that govern our universe. I am referring here to the realization that string theory gives us not a unique set of particles and forces but an infinite list out of which one came to be selected for our universe. We physicists have now to understand Darwin's lesson: The only way to understand how one out of a vast number of choices was made, which favors improbable structure, is that it is the result of evolution by natural selection.

Can this work? I showed that it might in 1992, in a theory of cosmological natural selection. This remains the only theory so far proposed of how our laws came to be selected that makes falsifiable predictions.

The idea that laws of nature evolved by natural selection is nothing new; it was anticipated by the philosopher Charles Sanders Peirce, who wrote in 1891:

> To suppose universal laws of nature capable of being apprehended by the mind and yet having no reason for their special forms, but standing inexplicable and irrational, is hardly a justifiable position. Uniformities are precisely the sort of facts that need to be accounted for. Law is par excellence the thing that wants a reason. Now the only possible way of accounting for the laws of nature, and for uniformity in general, is to suppose them results of evolution.

This idea remains dangerous, not only for what it has achieved but for what it implies for the future. For its implications have yet to be absorbed or understood, even by those who have come to believe this is the only way forward for science. For example, must there always be a deeper law, or meta-law, that governs the physical mechanisms by which a law evolves? And what about the fact that laws of physics are expressed in mathematics, which is usually thought of as encoding eternal truths? Can mathematics itself come to be seen as time-bound rather than as transcendent and eternal platonic truth?

I believe we will achieve clarity on these and other scary implications of the idea that all the regularities we observe, including those we have got used to calling laws, are the result of evolution by natural selection. And I believe that once this is achieved, Einstein and Darwin will be understood as partners in the greatest revolution yet in science, a revolution that taught us that the world in which we are embedded is nothing but an ever-evolving network of relationships.

The Multiverse

BRIAN GREENE

Brian Greene is a physicist and mathematician at Columbia University and the author, most recently, of *The Fabric of the Cosmos*.

The notion that there are universes beyond our own – the idea that we are but one member of a vast collection of universes called the multiverse – is highly speculative but both exciting and humbling. It's also an idea that suggests a radically new but inherently risky approach to certain scientific problems.

An essential working assumption in the sciences is that with adequate ingenuity, technical facility, and hard work, we can explain what we observe. The impressive progress made over the past few hundred years is testament to the apparent validity of this assumption. But if we are part of a multiverse, then our universe may have properties beyond traditional scientific explanation. Here's why:

Theoretical studies of the multiverse (within inflationary cosmology and string theory, for example) suggest that the detailed properties of the other universes may be significantly different from our own. In some, the particles making up matter may have different masses or electric charges; in others, the fundamental forces may differ in strength and even number from those we experience; in still others, the very structure of space and time may be unlike anything we've ever experienced.

In this context, the quest for fundamental explanations of particular properties of our universe – for example, the observed strengths of the nuclear and electromagnetic forces – takes on a very different character. The strengths of these forces may vary from universe to universe and thus it may simply be a matter of chance that in our universe these forces have the particular strengths they do. More intriguing, we can even imagine that in the other universes, where their strengths are different, conditions are not hospitable to our form of life. (With different force strengths, the processes giving rise to long-lived stars and stable planetary systems on which life can form and evolve can easily be disrupted.) In this setting, there would be no deep explanation for the observed force strengths. Instead, we would find ourselves living in a universe in which the forces have their familiar strengths simply because we couldn't survive in any of the others where the strengths were different.

If true, the idea of a multiverse would be a Copernican Revolution realized on a cosmic scale. It would be a rich and astounding upheaval but one with potentially hazardous consequences. Beyond the inherent difficulty in assessing its validity, when should we allow the multiverse framework to be invoked in lieu of a more traditional scientific explanation? Had this idea surfaced a hundred years ago, might researchers have chalked up various mysteries to how things just happen to be in our corner of the multiverse and not pressed on to discover all the wondrous science of the last century?

Thankfully that's not how the history of science played itself out – at least, not in our universe. But the point is manifest. While some mysteries may indeed reflect nothing more than the particular universe we find ourselves inhabiting within the multiverse, other mysteries are worth struggling with because they are the result of deep, underlying physical laws. The danger, if the multiverse idea takes root, is that researchers may too quickly give up the search for such underlying explanations. When faced with seemingly inexplicable observations, researchers may invoke the framework of the multiverse prematurely – proclaiming some phenomenon or other to merely reflect conditions in our own bubble universe and thereby failing to discover the deeper understanding that awaits us.

What Twentieth-Century Physics Says About the World Might Be True

CARLO ROVELLI

Carlo Rovelli is a professor of physics at the Centre de Physique Théorique de Luminy, Université de la Mediterranée, Marseille, and the author of *Quantum Gravity*.

There is a major dangerous scientific idea in contemporary physics, with a potential impact comparable to that made by Copernicus or Darwin. It is the idea that what the physics of the twentieth century says about the world might in fact be true.

Quantum mechanics must dramatically change our understanding of reality. If we take it seriously, we cannot, for instance, think that objects have a definite position. They have a position only when they interact with something else, and even then they have that position only with respect to that something else – they are still without position with respect to the rest of the world.

This is a change in our image of the world far more radical than that given by Copernicus, and a change in our way of thinking about ourselves far more consequential than that given by Darwin.

Still, few people take the quantum revolution seriously. The danger is exorcized by various strategies, for instance by saying something on the order of 'Well, quantum mechanics is relevant only for atoms and very small objects.' We still haven't fully recognized that the *world* is quantum mechanical, or accepted the immense conceptual revolution needed to make sense of this basic factual discovery about nature.

Another example is Einstein's relativity theory. Relativity makes clear that asking, 'What is happening right now on Andromeda?' is complete nonsense. There is no 'right now' elsewhere in the universe; nevertheless, we keep thinking of the universe as if there were an immense clock ticking away the instants, and we have a lot of difficulty adapting to the idea that a phrase like 'the present state of the universe right now' is physical nonsense.

In these cases, what we do is use concepts we have developed in our very special environment – which is characterized by low velocities and low energy – and we think of the world as if it were all just like that. We are like ants in a little garden with green grass and pebbles, who cannot apprehend any reality different from one of green grass and pebbles.

Many of today's audacious scientific speculations – about extra dimensions, the multiverse, and the like – are not only unsupported experimentally but are also quite often formulated within a worldview that has not even fully digested quantum mechanics and relativity!

It's a Matter of Time

PAUL STEINHARDT

Paul Steinhardt is the Albert Einstein Professor of Science at Princeton University.

For decades, the commonly held view among scientists has been that space and time first emerged about 14 billion years ago in a Big Bang. According to this picture, the cosmos transformed from a nearly uniform gas of elementary particles to its current complex hierarchy of structure, ranging from quarks to galaxy superclusters, through an evolutionary process governed by simple, universal physical laws. In the past few years, though, confidence in this point of view has been shaken as physicists have discovered finely tuned features of our universe that seem to defy natural explanation.

The prime culprit is the cosmological constant, which astronomers have measured to be exponentially smaller than naïve estimates would predict. On the one hand, it is crucial that the cosmological constant be so small, or else it would cause space to expand so rapidly that galaxies and stars would never form. On the other hand, no theoretical mechanism has been found within the standard Big Bang picture that would explain the tiny value.

Desperation has led to a dangerous idea: Perhaps we live in an anthropically selected universe. According to this view, we live in a multiverse (a multitude of universes) in which the cosmological constant varies randomly from one universe to the next. In most universes, the value is incompatible with the formation of galaxies, planets, and stars. The reason why our cosmological constant has the value it does is because it is one of the rare examples in which the value happens to lie in the narrow range compatible with life.

This is the ultimate example of 'unintelligent design': the multiverse tries every possibility with reckless abandon and only very rarely gets things 'right' – that is, consistent with everything we actually observe. It suggests that the creation of unimaginably enormous volumes of uninhabitable space is essential to obtain a few rare habitable spaces.

I consider this approach extremely dangerous for two reasons: First, it relies on complex assumptions about physical conditions far beyond the range of conceivable observation, so it is not scientifically verifiable. Second, it leads inevitably to a depressing end to science. What is the point of exploring further the randomly chosen physical properties in our tiny corner of the multiverse if most of the multiverse is so different? It is far too early to be so desperate. This is a dangerous idea I am simply unwilling to contemplate.

My own dangerous idea is more optimistic – but precarious, because it bucks the current trends in cosmological thinking. I believe that the finely tuned features may be naturally explained by supposing that our universe is much older than we have imagined. With more time, a new possibility emerges. The cosmological 'constant' may not be constant after all. Perhaps it is varying so slowly that it only appears to be constant. Originally it had the much larger value that we would naturally estimate, but the universe is so old that its value has had a chance to relax to the tiny value measured today. Furthermore, in several concrete examples one finds that the evolution of the cosmological constant slows down as its value approaches zero, so most of the history of the universe transpires when its value is tiny, just as we find today.

This idea that the cosmological constant is decreasing has been considered in the past. In fact, physically plausible slow-relaxation mechanisms have been identified. But the timing was thought to be impossible. If the cosmological constant decreases very slowly, it causes the expansion rate to accelerate too early and galaxies never form. If it decreases too quickly, the expansion rate never accelerates, which is inconsistent with recent observations. As long as the cosmological constant has only 14 billion years to evolve, there is no feasible solution.

But recently some cosmologists have been exploring the possibility that the universe is exponentially older. In this picture, the evolution of the universe is cyclic. The Big Bang is not the beginning of space and time but a sudden creation of hot matter and radiation that marks the transition from one period of expansion and cooling to the next cycle of evolution. Each cycle might last a trillion years, say. Fourteen billion years marks the time since the last infusion of matter and radiation, but this is brief compared to the total age of the universe. Each cycle lasts about a trillion years and the number of cycles in the past may have been ten to the googol power or more!

Then, using the aforementioned slow-relaxation mechanisms, it becomes possible that the cosmological constant decreases steadily from one cycle to the next. Since the number of cycles is likely to be enormous, there is enough time for the cosmological constant to shrink by an exponential factor, even though the decrease over the course of any one cycle is too small to be undetectable. Because the evolution slows down as the cosmological constant decreases, this is the period when most of the cycles take place. There is no multiverse and there is nothing special about our region of space – we live in a typical region at a typical time.

Remarkably, this idea is scientifically testable. The picture makes explicit predictions about the distribution of primordial gravitational waves and variations in temperature and density. Also, if the cosmological constant is evolving at the slow rate suggested, then ongoing attempts to detect a temporal variation should find no change. So, we may enjoy speculating now about which dangerous ideas we prefer, but ultimately it is nature that will decide if any of them are right. It is just a matter of time.

A Radical Re-evaluation of the Character of Time

PIET HUT

Piet Hut is a professor of astrophysics and interdisciplinary studies at the Institute for Advanced Study, in Princeton. He is the coauthor (with Douglas Heggie) of *The Gravitational Million-Body Problem: A Multidisciplinary Approach to Star Cluster Dynamics*.

Copernicus and Darwin took away our traditional place in the world and our traditional identity in the world. What traditional trait will be taken away from us next? My guess is that it will be the world itself. We see the first few steps in that direction in the physics, mathematics, and computer science of the twentieth century – from quantum mechanics to the results obtained by Gödel, Turing, and others. The ontologies of our world, concrete as well as abstract, have already started to melt away.

Copernicus upset the moral order by dissolving the strict distinction between heaven and earth. Darwin did the same by dissolving the strict distinction between humans and other animals. Could the next step be the dissolution of the strict distinction between reality and fiction? For this to be shocking, it has to come in a scientifically respectable way, as a precise and inescapable conclusion; it should have the technical strength of a body of knowledge like quantum mechanics, as opposed to collections of opinions on the level of cultural relativism.

Perhaps a radical re-evaluation of the character of time will do it. In everyday experience time flows, and we flow with it. In classical physics, time is frozen as part of a frozen spacetime picture. And there is as yet no agreed-upon interpretation of time in quantum mechanics. What if a future scientific understanding of time were to show all previous pictures to be wrong and demonstrate that past, future, and even the present do not exist? That stories woven around our individual personal history and future are all just wrong? Now, that would be a dangerous idea!

It's OK Not to Know Everything

MARCELO GLEISER

Marcelo Gleiser is a professor of physics and astronomy at Dartmouth College and the author of *The Dancing Universe: From Creation Myths to the Big Bang*.

There have been many times when I asked myself whether we scientists, especially those seeking to answer ultimate kinds of questions such as the origin of the universe, are not beating the wrong drum. By trying to answer such questions as the origin of everything, we assume we can. We plow ahead, proposing tentative models that join general relativity and quantum mechanics where the universe pops out of nothing, no energy required: All is due to a random quantum fluctuation. To this, we add the randomness of fundamental constants, saying that their values are due to accident; other universes may well have other values of the charge and mass of the electron and thus completely different properties. So our universe becomes this very special place where things conspire to produce galaxies, stars, planets, and life.

What if all this is bogus? What if we look at science as a narrative, a description of the world that has limitations based on its structure? The constants of nature are the letters of the alphabet, the laws of nature the grammar rules, and we build these descriptions through the guiding hand of the so-called scientific method. Period. To say things are this way because otherwise we wouldn't be here to ask the questions is to miss the point altogether. Things are this way because this is the story we humans tell based on the way we see the world and explain it.

If we take this view to the extreme, it means that we will never be able to answer the question of the origin of the universe, since it implicitly assumes that science can explain itself. We can build any cool and creative models we want, using any marriage of quantum mechanics and relativity, but we still won't understand why these laws are the laws and not others. In a sense, this means that our science is our science and not something universally true, as many believe. This is not bad at all, given what we can do with it, but it does place limits on knowledge. Which may also not be a bad thing. It's OK not to know everything. It doesn't make science weaker. Only more human.

The End of Insight

STEVEN STROGATZ

Steven Strogatz is an applied mathematician at Cornell University and the author of *Sync: How Order Emerges from Chaos in the Universe, Nature, and Daily Life.*

I worry that insight is becoming impossible, at least at the frontiers of mathematics. Even when we're able to figure out what's true or false, we're less and less able to understand why.

An argument along these lines was recently given by Brian Davies in the *Notices of the American Mathematical Society*. He mentions, for example, that the four-color-map theorem in topology was proved in 1976 with the help of computers, which exhaustively checked a huge but finite number of possibilities. No human mathematician could ever verify all the intermediate steps in this brutal proof, and even if someone claimed to, should we trust them? To this day, no one has come up with a more elegant, insightful proof. So we're left in the unsettling position of knowing that the four-color theorem is true but still not knowing why.

Similarly important but unsatisfying proofs have appeared in group theory (in the classification of finite simple groups, roughly akin to the periodic table for chemical elements) and in geometry (in the problem of how to pack spheres so that they fill space most efficiently, a puzzle that goes back to Kepler in the sixteenth century and arises today in coding theory for telecommunications).

In my own field of complex-systems theory, Stephen Wolfram has emphasized that there are simple computer programs known as cellular automata whose dynamics can be so inscrutable that there's no way to predict how they'll behave. The best you can do is simulate them on the computer, sit back, and watch how they unfold. Observation replaces insight. Mathematics becomes a spectator sport.

If this is happening in mathematics, the supposed pinnacle of human reasoning, it seems likely to afflict us in science, too – first in physics and later in biology and the social sciences (where we're not even sure what's true, let alone why).

When the End of Insight comes, the nature of explanation in science will change forever. We'll be stuck in an age of authoritarianism, except it will no longer be coming from politics or religious dogma but from science itself.

When Will the Internet Become Aware of Itself?

TERRENCE SEJNOWSKI

Terrence Sejnowski is a computational neuroscientist at the Howard Hughes Medical Institute and coauthor (with Steven R. Quartz) of *Liars, Lovers, and Heroes: What the New Brain Science Reveals About How We Become Who We Are.*

I never thought I would become omniscient during my lifetime, but as Google continues to improve and on-line information continues to expand, I have achieved omniscience for all practical purposes. The Internet has created a global marketplace for ideas and products, making it possible for individuals in the far corners of the world to automatically connect directly to one another. The Internet has achieved these capabilities by growing exponentially in total communications bandwidth. How does the communications power of the Internet compare with that of the cerebral cortex, the most interconnected part of our brains?

Cortical connections are expensive, because they take up volume and cost energy to send information, in the form of spikes along axons. About 44 percent of the cortical volume in humans is taken up with long-range connections, called the white matter. Interestingly, the thickness of gray matter, just a few millimeters, is nearly constant in mammals that range in brain volume over 5 orders of magnitude, and the volume of the white matter scales approximately as the 4/3 power of the volume of the gray matter. The larger the brain, the larger the fraction of resources devoted to communications compared to computation.

However, the global connectivity in the cerebral cortex is extremely sparse: The probability of any two cortical neurons having a direct connection is around 1 in 100 for neurons in a vertical column 1 mm in diameter but only 1 in 1,000,000 for more distant neurons. Thus, only a small fraction of the computation that occurs locally can be reported to other areas, through a small fraction of the cells that connect distant cortical areas.

Despite the sparseness of cortical connectivity, the potential bandwidth of all of the neurons in the human cortex is approximately 1 terabit per second, comparable to the total world backbone capacity of the Internet. However, this capacity is never achieved by the brain in practice, because only a fraction of cortical neurons have a high rate of firing at any given time. Recent work by the neurobiologist Simon Laughlin suggests that another physical constraint – energy – limits the brain's ability to harness its potential bandwidth.

The cerebral cortex also has a massive amount of memory. There are approximately 10^9 synapses between neurons under every square millimeter of cortex, or about 10^{11} synapses overall. Assuming around a byte of storage capacity at each synapse (including dynamic as well as static properties), this comes to a total of 10^{15} bits of storage. This is comparable to the amount of data on the entire Internet. Google can store this in terabyte disk arrays and has hundreds of thousands of computers simultaneously sifting through it.

Thus, the Internet and our ability to search it are within

reach of the limits of the raw storage and communications capacity of the human brain, and should exceed it by 2015.

The biophysicist Leo van Hemmen and I recently asked twenty-three neuroscientists to think about what we don't yet know about the brain and propose a question so fundamental and difficult that it could take a century to solve – in the spirit of David Hilbert's twenty-three problems in mathematics. Christof Koch and Francis Crick speculated that the key to understanding consciousness was global communication: How do neurons in the diverse parts of the brain manage to coordinate despite the limited connectivity? Sometimes, the communication gets crossed, and V. S. Ramachandran and Edward Hubbard asked whether synesthetes – rare individuals who experience crossover in sensory perception, such as hearing colors, seeing sounds, and tasting tactile sensations – might give us clues to how the brain evolved.

There is growing evidence that the flow of information between parts of the cortex is regulated by the degree of synchrony of the spikes within populations of cells that represent perceptual states. Robert Desimone and his colleagues have examined the effects of attention on cortical neurons in awake, behaving monkeys and found the coherence between the spikes of single neurons in the visual cortex and local field potentials in the gamma band, 30-80 Hz, increased when the covert attention of a monkey was directed toward a stimulus in the receptive field of the neuron. The coherence also selectively increased when a monkey searched for a target with a cued color or shape amid a large number of distracters. The increase in coherence means that neurons representing the stimuli with the cued feature would have greater impact on target neurons, making them more salient.

The link between attention and spike-field coherence raises a number of interesting questions. How does top-down input from the prefrontal cortex regulate the coherence of neurons in other parts of the cortex through feedback connections? How is the rapidity of the shifts in coherence achieved? Experiments on neurons in cortical slices suggest that inhibitory interneurons are connected to each other in networks and are responsible for gamma oscillations. Researchers in my laboratory have used computational models to show that excitatory inputs can rapidly synchronize a subset of the inhibitory neurons that are in competition with other inhibitory networks. Inhibitory neurons, long thought to merely block activity, are highly effective in synchronizing neurons in a local column already firing in response to a stimulus.

The oscillatory activity that is thought to synchronize neurons in different parts of the cortex occurs in brief bursts, typically lasting for only a few hundred milliseconds. Thus, it is possible that there is a packet structure for long distance communication in the cortex, similar to the packets that are used to communicate on the Internet, though with quite different protocols. The first electrical signals recorded from the brain in 1875 by Richard Caton were oscillatory signals that changed in amplitude and frequency with the state of alertness. The function of these oscillations remains a mystery, but it would be remarkable if it were to be discovered that these signals held the secrets to the brain's global communications network.

Since its inception in 1969, the Internet has been scaled up to a size unimagined even by its inventors, in contrast to most engineered systems, which fall apart when they are pushed beyond their design limits. In part, the Internet achieves this scalability because it can regulate itself, deciding on the best routes to send packets depending on traffic conditions. Like the brain, the Internet has circadian rhythms that follow the sun as the planet rotates under it. The growth of the Internet over the last several decades more closely resembles biological evolution than engineering.

How would we know if the Internet were to become aware of itself? The problem is that we don't even know if some of our fellow creatures on this planet are self-aware. For all we know, the Internet is already aware of itself.

Democratizing Access to the Means of Invention

NEIL GERSHENFELD

Neil Gershenfeld is a physicist and the director of the Center for Bits and Atoms at MIT. He is the author of *Fab: The Coming Revolution on Your Desktop — From Personal Computers to Personal Fabrication.*

The elite temples of research (of the kind I've happily spent my career in) may be becoming intellectual dinosaurs as a result of the digitization and personalization of fabrication.

Today, with about $20,000 in equipment, it's possible to make and measure things from microns and microseconds on up, and that boundary is quickly receding. When I came to MIT, that was hard to do. If it's no longer necessary to go to MIT for its facilities, then surely the intellectual community is its real resource? But my colleagues and I are always either traveling or overscheduled; the best way for us to see one another is to go somewhere else. Like many people, my closest collaborators are distributed around the world.

The ultimate consequence of the digitization of communications, then computation, and now fabrication is to democratize access to the means of invention. The third world can skip over the first and second cultures and go right to developing a third culture. Rather than today's model of researchers researching for researchees, the result of all that discovery has been to enable a planet of creators rather than consumers.

Mind Is a Universally Distributed Quality

RUDY RUCKER

Rudy Rucker is a mathematician, computer scientist, cyberpunk pioneer, and novelist. He is the author, most recently, of *The Lifebox, the Seashell, and the Soul: What Gnarly Computation Taught Me About Ultimate Reality, the Meaning of Life, and How to Be Happy*.

Panpsychism. Each object has a mind. Stars, hills, chairs, rocks, scraps of paper, flakes of skin, molecules – each of them possesses the same inner glow as a human, each of them has singular inner experiences and sensations.

I'm quite comfortable with the notion that everything is a computation. But what to do about my sense that there's something numinous about my inner experience? Panpsychism represents a nonanthropocentric way out: Mind is a universally distributed quality.

Yes, the workings of a human brain are a deterministic computation that could be emulated by any universal computer. And, yes, I sense more to my mental phenomena than the rule-bound exfoliation of reactions to inputs: This residue is the inner light, the raw sensation of existence. But, no, that inner glow is not the exclusive birthright of humans, nor is it solely limited to biological organisms.

Note that panpsychism needn't say that the universe is just one mind. We can also say that each object has an individual mind. One way to visualize the distinction between the many minds and the one mind is to think of the world as a stained glass window with light shining through each pane. The world's physical structures break the undivided cosmic mind into a myriad of small minds, one in each object.

The minds of panpsychism can exist at various levels. As well as having its own individuality, a person's mind would also be, for instance, a hive mind, based on the minds of the body's cells and the minds of the body's elementary particles.

Do the panpsychic minds have any physical correlates? On the one hand, it could be that the mind is some substance that accumulates near ordinary matter — dark matter or dark energy are good candidates. On the other hand, mind might simply be matter viewed in a special fashion: matter experienced from the inside. Let me mention three specific physical correlates that have been proposed for the mind.

Some have argued that the experience of mind results when a superposed quantum state collapses into a pure state. It's an alluring metaphor, but as a universal automatist I'm of the opinion that quantum mechanics is a stopgap theory, destined to give way to a fully deterministic theory based on some digital precursor of spacetime.

David Skrbina, author of the clear and comprehensive book *Panpsychism in the West*, suggests that we might think of a physical system as determining a moving point in a multidimensional phase space that has an axis for each of the system's measurable properties. He feels that this dynamic point represents the sense of unity characteristic of a mind.

As a variation on this theme, let me point out that from the universal automatist standpoint, every physical system can be thought of as embodying a computation. And the majority of nonsimple systems embody universal computations, capable of emulating any other system at all. It could be that having a mind is in some sense equivalent to being capable of universal computation.

A side remark: Even such very simple systems as a single electron may be capable of universal computation, if afforded a steady stream of structured input. Think of an electron in an oscillating field – and, by analogy, think of a person listening to music or reading an essay.

Might panpsychism be a distinction without a difference? Suppose we identify the numinous mind with quantum collapse, with chaotic dynamics, or with universal computation. What is added by claiming that these aspects of reality are like minds?

I think empathy can supply an experiential confirmation of panpsychism's reality. Just as I'm sure that I myself have a mind, I can come to believe the same of another human with whom I'm in contact – whether face to face or via their creative work. And with a bit of effort, I can identify with objects as well; I can see the objects in the room around me as glowing with inner light. This is a pleasant sensation; one feels less alone.

Could there ever be a critical experiment to test if panpsychism is really true? Suppose telepathy were to become possible, perhaps by entangling a person's mental states with another system's states. And then suppose that instead of telepathically contacting another person, I were to contact a rock. At this point, panpsychism would be proved.

I still haven't said anything about why panpsychism is a dangerous idea. Panpsychism, like other forms of higher consciousness, is dangerous to business as usual. If my old car has the same kind of mind as a new one, I'm less impelled to help the economy by buying a new vehicle. If the rocks and plants on my property have minds, I feel more respect for them in their natural state. If I feel myself among friends in the universe, I'm less likely to overwork myself to earn more cash. If my body will have a mind even after I'm dead, then death matters less to me and it's harder for the government to cow me into submission.

The Forbidden Fruit Intuition

THOMAS METZINGER

Thomas Metzinger is a professor of philosophy at Johannes Gutenberg-Universität, Mainz, and the author of *Being No One: The Self-Model Theory of Subjectivity*.

We would all like to believe that intellectual honesty is not only an expression of but also good for your mental health. My dangerous question is whether one can be intellectually honest about the issue of free will and preserve one's mental health at the same time. Behind this question lies what I call the Forbidden Fruit intuition: Is there a set of questions that are dangerous not on grounds of ideology or political correctness but because the most obvious answers to them could ultimately make our conscious self-models disintegrate? Can one really believe in determinism without going insane?

For middle-sized objects at 37°C, like the human brain and the human body, determinism is obviously true. The next state of the physical universe is always determined by the previous state. And given a certain brain-state plus an environment, you could never have acted otherwise; a surprisingly large majority of experts in the free-will debate today accept this obvious fact. Although your future is open, this probably also means that for every single future thought you will have and for every single decision you will make, it is true that it was determined by your previous brain state.

As a scientifically well-informed person, you believe in this theory – you endorse it. As an open-minded person, you find that you are also interested in modern philosophy of mind, and you might hear a story much like the following one: Yes, you are a physically determined system. But this is not a big problem, because under certain conditions we may still continue to say that you are 'free.' All that matters is that your actions are caused by the right kinds of brain processes and that they originate in you. A physically determined system can well be sensitive to reasons and to rational arguments, to moral considerations, to questions of value and ethics, as long as all of this is appropriately wired into its brain. You can be rational, and you can be moral, as long as your brain is physically determined in the right way.

You like this basic idea: Physical determinism is compatible with being a free agent. You endorse a materialist philosophy of freedom as well. An intellectually honest person open to empirical data, you simply believe that something along these lines must be true.

Now you try to feel that it is true. You try to consciously experience the fact that at any given moment of your life you could not have acted otherwise. You try to experience the fact that even your thoughts, however rational and moral, are predetermined – by something unconscious, by something you cannot see. And in doing so, you start fooling around with the conscious self-model that Mother Nature evolved for you with so much care and precision over millions of years. You are scratching at the user surface of your own brain, tweaking the mouse pointer, introspectively trying to penetrate into the operating system, attempting to make the invisible visible. You are challenging the integrity of your phenomenal self by trying to integrate your new beliefs, the neuroscientific image of a human being, with your most intimate, inner way of experiencing yourself. How does it feel?

I think that the irritation and deep sense of resentment surrounding public debates on the freedom of the will actually has nothing much to do with the actual options on the table. It has to do with the (perfectly sensible) intuition that our currently obvious answer will not only be emotionally disturbing but ultimately impossible to integrate into our conscious self-models.

Or our societies: The robust conscious experience of free will also is a social institution, because the attribution of accountability, responsibility, and so on are the building blocks of modern open societies. And the currently obvious answer might be interpreted by many as having clearly antidemocratic implications: Making a complex society work implies controlling the behavior of millions of people; if individual human beings can control their own behavior to a much lesser degree than we have thought in the past, if *bottom-up* doesn't work, then it becomes tempting to control it *top-down*, by the state. And this is the second way in which enlightenment could devour its own children. Yes, free will truly is a dangerous question, but for different reasons than most people think.

The Posterior Probability of Any Particular God Is Pretty Small

PHILIP W. ANDERSON

Philip W. Anderson is a physicist at Princeton University and the author of *Concepts in Solids*.

Isn't God very improbable? You can't, in any logical system I can understand, disprove the existence of God – or prove it for that matter. But in the probability calculus I use, he is very improbable.

There are a number of ways of making a formal probability theory which incorporate Ockham's razor (the principle that one must not multiply hypotheses unnecessarily). Two of them are Bayesian probability theory and minimum entropy. If you have been taking data on something and the data are reasonably close to a straight line, these methods give us a definable procedure by which you can estimate the probability that the straight line is correct – not the polynomial that has as many parameters as there are points, or some intermediate complex curve. Ockham's razor is expressed mathematically as the fact that there is a factor in the probability derived for a given hypothesis that decreases exponentially in the number n of parameters that describe your hypothesis. It is the inverse of the

PHILIP W. ANDERSON 149

volume of parameter space. People who are trying to prove the existence of ESP (extrasensory perception), for example, abominate Bayesianism because it strongly favors the 'null hypothesis' and beats them every time.

Well, now, imagine how big the parameter space is for God. He could have a long gray beard or not; be benevolent or malicious in a lot of different ways and over a wide range of values; have a variety of views on abortion and contraception; like or abominate human images; like or abominate music – and the range of dietary prejudices he has been credited with is as long as your arm. There is the Heaven-Hell dimension, the one-versus-three question, and I haven't even mentioned polytheism. I think there are certainly as many parameters as sects, or more. If there is even a sliver of prior probability for the null hypothesis, the posterior probability of any particular God is pretty small.

Science Must Destroy Religion

SAM HARRIS

Sam Harris is the author of *The End of Faith* and *Letter to a Christian Nation*. He is completing a doctorate in neuroscience, using functional magnetic resonance imaging (fMRI) to study the neural basis of belief, disbelief, and uncertainty.

Most people believe that the creator of the universe wrote (or dictated) one of their books. Unfortunately, there are many books that pretend to divine authorship, and each makes incompatible claims about how we all must live. Despite the ecumenical efforts of many well-intentioned people, these irreconcilable religious commitments still inspire an appalling amount of human conflict.

In response to this situation, most sensible people advocate something called 'religious tolerance.' While religious tolerance is surely better than religious war, tolerance is not without its liabilities. Our fear of provoking religious hatred has rendered us incapable of criticizing ideas that are now patently absurd and increasingly maladaptive. It has also obliged us to lie to ourselves – repeatedly and at the highest levels – about the compatibility between religious faith and scientific rationality.

The conflict between religion and science is inherent and

(very nearly) zero-sum. The success of science often comes at the expense of religious dogma; the maintenance of religious dogma always comes at the expense of science. It is time we conceded a basic fact of human discourse: Either people have good reasons for what they believe, or they do not. When they have good reasons, their beliefs contribute to our growing understanding of the world. We need not distinguish between 'hard' and 'soft' sciences here, or between science and other evidence-based disciplines, like history. There happen to be very good reasons to believe that the Japanese bombed Pearl Harbor on December 7, 1941. Consequently, the idea that the Egyptians actually did it lacks credibility. Every sane human being recognizes that to rely merely on faith to decide specific questions of historical fact would be both idiotic and grotesque – that is, until the conversation turns to the origin of books like the Bible and the Koran, to the resurrection of Jesus, to Muhammad's conversation with the angel Gabriel, or to any of the other hallowed travesties that still crowd the altar of human ignorance.

Science, in the broadest sense, includes all reasonable claims to knowledge about ourselves and the world. If there were good reasons to believe that Jesus was born of a virgin, or that Muhammad flew to Heaven on a winged horse, these beliefs would necessarily form part of our rational description of the universe. Faith is nothing more than the license that religious people give one another to believe such propositions when reasons fail. The difference between science and religion is the difference between a willingness to dispassionately consider new evidence and new arguments and a passionate unwillingness to do so. The distinction could not be more obvious, or more consequential, and yet it is everywhere elided, even in the ivory tower.

Religion is fast growing incompatible with the emergence of a global civil society. Religious faith – faith that there is a God who cares what name he is called, that one of our books is infallible, that Jesus is coming back to Earth to judge the living and the dead, that Muslim martyrs go straight to Paradise, and so on – is on the wrong side of an escalating war of ideas. The difference between science and religion is the difference between a genuine openness to fruits of human inquiry in the twenty-first century and a premature closure to such inquiry as a matter of principle. I believe that the antagonism between reason and faith will only grow more pervasive and intractable in the coming years. Iron Age beliefs – about God, the soul, sin, free will, and so on – continue to impede medical research and distort public policy. The possibility that we could elect a U.S. president who takes biblical prophecy seriously is real and terrifying; the likelihood that we will one day confront Islamists armed with nuclear or biological weapons is also terrifying and growing more probable by the day. We are doing very little, at the level of our intellectual discourse, to prevent such possibilities.

In the spirit of religious tolerance, most scientists are keeping silent when they should be blasting the hideous fantasies of a prior age with all the facts at their disposal.

To win this war of ideas, scientists and other rational people will need to find new ways of talking about ethics and spiritual experience. The distinction between science and religion is not a matter of excluding our ethical intuitions and nonordinary states of consciousness from our conversation about the world; it is a matter of our being rigorous about what is reasonable to conclude on their basis. We must find ways of meeting our emotional needs that do not require the abject embrace of the preposterous. We must learn to invoke the power of ritual and to mark those transitions in every human life that demand profundity – birth, marriage, death – without lying to ourselves about the nature of reality.

I am hopeful that the necessary transformation in our thinking will come about as our scientific understanding of ourselves matures. When we find reliable ways to make human beings more loving, less fearful, and genuinely enraptured by the fact of our appearance in the cosmos, we will have no need for divisive religious myths. Only then will the practice of raising our children to believe that they are Christian, Jewish, Muslim, or Hindu be broadly recognized as the ludicrous obscenity that it is. And only then will we stand a chance of healing the deepest and most dangerous fractures in our world.

The Self Is a Conceptual Chimera

JOHN ALLEN PAULOS

John Allen Paulos is a professor of mathematics at Temple University and the author of *Innumeracy*.

Doubt that a supernatural being exists is banal, but the more radical doubt that we exist, at least as anything more than nominal, marginally integrated entities having convenient labels like 'Myrtle' and 'Oscar,' is my candidate for a dangerous idea. This is, of course, the philosopher David Hume's idea – and the Buddha's as well: that the self is an ever changing collection of beliefs, perceptions, and attitudes, that it is not an essential and persistent entity but a conceptual chimera. If this belief ever became widely and viscerally felt throughout a society – whether because of advances in neurobiology, cognitive science, philosophical insights, or whatever – its effects on that society would be incalculable. (Or so this particular assemblage of beliefs, perceptions, and attitudes sometimes thinks.)

The Greatest Story Ever Told

CAROLYN C. PORCO

Carolyn C. Porco is a planetary scientist, leader of the Cassini Imaging Team, director of CICLOPS (Cassini Imaging Laboratory for Central Operations), and adjunct professor at the University of Arizona.

The confrontation between science and formal religion will come to an end when the role played by science in the lives of all people is the same as that played by religion today.

And just what is that role?

At the heart of every scientific inquiry is a deep spiritual quest – to grasp, to know, to feel connected through an understanding of the secrets of the natural world, to have a sense of one's part in the greater whole. It is this inchoate desire for connection to something greater and immortal, the need for elucidation of the meaning of the 'self,' that motivates the religious to belief in a higher intelligence. It is the allure of a bigger agency – outside the self but also involving, protecting, and celebrating the purpose of the self – that is the great attractor. Every culture has religion. It manifestly satisfies a human need.

But the same spiritual fulfillment and connection can be found in the revelations of science. From energy to matter, from fundamental particles to DNA, from microbes to *Homo sapiens*, from the singularity of the Big Bang to the immensity of the universe . . . ours is the greatest story ever told. We scientists have the drama, the plot, the icons, the spectacles, the 'miracles,' the magnificence, and even the special effects. We inspire awe. We evoke wonder. And we don't have one god, we have many. We find gods in the nucleus of every atom, in the structure of spacetime, in the counterintuitive mechanisms of electromagneticsm. What richness! What consummate beauty!

We even exalt the 'self.' Our script requires a broadening of the usual definition, but we, too, offer hope for everlasting existence. The self that is the particular, networked set of connections of matter comprising our mortal bodies will one day die, of course. But the self that is the sum of each separate individual condensate in us of energy-turned-matter is already ancient and will live forever. Each fundamental particle may one day return to energy or from there revert back to matter. But in one form or another, it will not cease. In this sense, we and all around us are eternal, immortal, and profoundly connected. We don't have one soul; we have trillions upon trillions of them.

These are reasons enough for jubilation – for riotous, unrestrained, exuberant merrymaking.

So what are we missing?

Ceremony.

We lack ceremony. We lack ritual. We lack the initiation of baptism, the brotherhood of communal worship.

We have no loving ministers, guiding and teaching the flocks in the ways of the 'gods.' We have no fervent missionaries, no loyal apostles. We lack the all-inclusive ecumenical embrace, the extended invitation to the masses. Alienation does not warm the heart; communion does.

But what if . . . ? What if we appropriated the craft, the artistry, the methods of formal religion to get the message across? Imagine Einstein's Witnesses going door-to-door or TV evangelists passionately declaiming the beauty of evolution.

Imagine a Church of Latter Day Scientists, where believers could gather. Imagine congregations raising their voices in tribute to gravity, the force that binds us all to the earth and the earth to the sun and the sun to the Milky Way. Or rejoicing in the nuclear force that makes possible the sunlight of our star and the starlight of distant suns. And can't you just hear the hymns sung to the antiquity of the universe, its abiding laws, and the Heaven above that we will all one day inhabit together, commingled, spread out like a nebula against a diamond sky?

One day the sites we hold most sacred just might be the astronomical observatories, the particle accelerators, the university research installations, and other laboratories where the high priests of science – the biologists, the physicists, the astronomers, the chemists – engage in the noble pursuit of uncovering the workings of nature. And today's museums, exposition halls, and planetaria may become tomorrow's houses of worship, where these revealed truths and the wonder of our interconnectedness with the cosmos are glorified in song by the devout and the soulful.

'Hallelujah!' they will sing. 'May the force be with you!'

Science As Just Another Religion

JORDAN POLLACK

Jordan Pollack directs a research laboratory in dynamical and evolutionary machine organization at Brandeis University.

We scientists like to think that our way of knowing is special. Instead of holding beliefs based on faith in invisible omniscient deities or parchments transcribed from oral cultures, we use the scientific method to discover and know. Truth may be eternal, but human knowledge of that truth evolves over time, as new questions are asked, data are recorded, hypotheses are tested, and replication and refutation mechanisms correct the record.

So it is a very dangerous idea to consider science as just another religion. It's not my idea but one I noticed growing in a set of Lakovian frames within the memesphere.

One of the frames is that scientists are doom-and-gloom prophets. At a recent popular technology conference, a parade of speakers emphasized the threats of global warming, sea level rising by eighteen feet and destroying cities, more category 5 hurricanes, and so on. It was quite a reversal from the positivistic techno-utopian promises of miraculous advances in medicine, computers, and weaponry that allowed science to bloom in the late twentieth century. A friend pointed out that in the days before PowerPoint these scientists might be wearing sandwich-board signs proclaiming, 'The End Is Near!'

Another element in the framing of science as a religion is the response to evidence-based policy. Scientists who do take political stands on 'moral' issues such as stem-cell research, capital punishment, nuclear weapons, global warming, and the like can be sidelined as atheists, humanists, or agnostics who have no moral or ethical standing outside their narrow specialty – as compared with, say, televangelists.

A third, and the most nefarious, frame casts theory as one opinion among others that should also be represented out of fairness or tolerance. This is the subterfuge used by the intelligent-design creationists.

We may believe in the separation of church and state, but that firewall has fallen. Science and reason are losing political battles to superstition and ignorance. Politics works by rewarding friends and punishing enemies, and while our individual votes may be private, exit polls have proved that science didn't vote for the incumbent.

There seem to be three choices put forward: reject, accommodate, or embrace.

One path is to go on an attack on religion in the public sphere. In his book *The End of Faith*, Sam Harris points out that humoring people who believe in God is like humoring people who believe that 'a diamond the size of a refrigerator' is buried in their backyard. There is a fine line between pushing God out of our public institutions and repeating religious intolerance of regimes past.

A second is to embrace faith-based science. Since, from the perspective of government, research is just another special interest feeding at the public trough, we should change our model to be more accommodating to political reality. Research is already sold like highway construction projects, with a linear accelerator for your state and a supercomputer center for mine, all done through direct appropriations. All that needs to change is the justifications for such spending.

How would faith-based science work? Well, physics could sing the psalm that perpetual motion would solve the energy crisis, thereby triggering a $500 billion program in free energy machines. (Of course, God is on our side to repeal the second law of thermodynamics!) Astronomy could embrace astrology and do grassroots PR with daily horoscopes to gain mass support for a new space program. In fact, an antigravity initiative could pass today if it was spun as a repeal of the 'heaviness tax.' Using the renaming principle, the SETI (search for extraterrestrial intelligence) program can be brought back to life as the 'Search for God' project.

Finally, the third idea is actually to *embrace* this dangerous idea and organize a new open-source spiritual and moral movement. I think a new, greener religion, based on faith in the Gaia hypothesis and an eleventh commandment to 'Protect the earth' could catch on, especially if welcoming to existing communities of faith. Such a movement could be a new pulpit from which the evidence-based silent majority can speak with both moral force and evangelical fervor about issues critical to the future of our planet.

This Is All There Is

ROBERT R. PROVINE

Robert R. Provine is a professor of psychology and assistant director of the Neuroscience Program at the University of Maryland Baltimore County. He is the author of *Laughter: A Scientific Investigation.*

The empirically testable idea that the here and now is all there is and that life begins at birth and ends at death is so dangerous that it has cost the lives of millions and threatens the future of civilization. The danger comes not from the idea itself but from its opponents, those religious leaders and followers who ruthlessly advocate and defend their empirically improbable afterlife and man-in-the-sky cosmological perspectives.

Their vigor is understandable. What better theological franchise is there than the promise of everlasting life, with deluxe trimmings? Religious followers must invest now with their blood and sweat, with the big payoff not due until the afterlife. Postmortem rewards cost theologians nothing – I'll match your heavenly choir and raise you seventy-two virgins.

Some franchise! This is even better than the medical profession, a calling with higher overhead that has gained control of birth, death, and pain. Whether the religious brand is Christianity or Islam, the warring continues, with a terrible fate reserved for heretics who threaten the franchise from within. Worse may be in store for those who totally reject the man-in-the-sky premise and its afterlife trappings. All of this trouble over accepting what our senses tell us – that this is all there is.

Resolution of religious conflict is impossible, because there is no empirical test of the ghostly, and many theologians prey, intentionally or not, on the fears, superstitions, irrationality, and herd tendencies that are our species' neurobehavioral endowment. Religious fundamentalism inflames conflict and prevents solution; the more extreme and irrational one's position, the stronger one's faith. When you are in possession of absolute truth, compromise is not an option.

Resolution of conflicts between religions and associated cultures is less likely to come from compromise than from the pursuit of superordinate goals – common, overarching, objectives that extend across nations and cultures and direct our competitive spirit to further the health, well-being, and nobility of everyone. Public health and science provide such unifying goals. I offer two examples.

Health Initiative. A program that improves the health of all people, especially those in developing nations, may find broad support, especially with the growing awareness of global culture and the looming specter of a pandemic. Public health programs bridge religious, political, and cultural divides. No one wants to see their children die. Conflicts fall away when cooperation offers a better life for all concerned. This is also the most effective antiterrorism strategy, although one probably unpopular both with the military-industrial complex and terrorist agitators.

Space Initiative. Space exploration expands our cosmos and increases our appreciation of life on Earth and its finite resources. Space exploration is one of our species' greatest achievements. Its pursuit is a goal of sufficient grandeur to unite people of all nations. This is all there is. The sooner we accept this dangerous idea, the sooner we can get on with the essential task of making the most of our lives on this planet.

A Science of the Divine?

STEPHEN M. KOSSLYN

Stephen M. Kosslyn is a professor of psychology at Harvard University and an associate psychologist in the Department of Neurology at the Massachusetts General Hospital. He is the author (with Olivier Koenig) of *Wet Mind: The New Cognitive Neuroscience*.

Here's an idea that many academics may find unsettling and dangerous: God exists. And here's another that many religious people may find unsettling and dangerous: God is not supernatural but rather part of the natural order.

Simply stating these ideas in the same breath invites them to scrape against each other, and sparks begin to fly. To avoid such conflict, Stephen Jay Gould famously argued that we should separate religion and science, treating them as distinct 'magisteria.' But science leads many of us to try to understand all that we encounter as existing in a single grand and glorious overarching framework. In this spirit, let me try to suggest one way in which the idea of a supreme being can fit into a scientific worldview. I offer the following not to advocate but simply to illustrate one (certainly not the only) way that the concept of God can be approached scientifically.

First, here's the specific conception of God I want to explore:

God is a supreme being that transcends space and time, permeates our world but also stands outside of it, and can intervene in our daily lives (partly in response to prayer).

A way to begin to think about this conception of the divine rests on three ideas:

(1) *Emergent properties.* There are many examples in science in which aggregates produce an entity that has properties that cannot be predicted entirely from the elements themselves. Neurons in large numbers produce minds; minds in large numbers produce economic, political, and social systems.

(2) *Downward causality.* Events at higher levels – levels where emergent properties become evident – can in turn feed back and affect events at lower levels. For example, chronic stress, a mental event, can cause parts of the brain to become smaller. Similarly, an economic depression or the results of an election affect the lives of the individuals who live in that society.

(3) *The Ultimate Superset.* The Ultimate Superset (superordinate set) of all living things may have an equivalent status to an economy or culture. It has properties that emerge from the interactions of living things and groups of living things and in turn can feed back to affect those things and groups.

Can we conceive of God as an emergent property of all living things that can in turn affect its constituents? Here are some ways in which this idea is consistent with the nature of God, as outlined at the outset:

This emergent entity is transcendent, in the sense that it exists in no specific place or time. Like a culture or an economy, God is nowhere, although the constituent elements occupy specific places. As for transcending time, consider this analogy: Imagine that 1/100th of the neurons in your brain were replaced every hour, and each old neuron programmed a new one so that the old one's functionality was preserved. After a hundred hours, your brain would be an entirely new organ, but your mind would continue to exist as it had been before. Similarly, as each citizen dies and is replaced by a child, the culture continues to exist (and can grow and develop, with a life of its own). So too with God. In the story of Jacob's ladder, Jacob realizes, 'Surely the *Lord* is in this place, and I knew it not.' (Genesis 28:16. I interpret this story as illustrating that God is everywhere but nowhere. The Ultimate Superset permeates our world but also stands outside of (or, more specifically, above) it.

The Ultimate Superset can affect our individual lives. Another analogy: Say that geese flying south for the winter have rather unreliable magnetic field detectors in their brains. However, there's a rule built into their brains that leads them to try to stay near their fellows as they fly. The flock as a whole would navigate far better than any individual bird, because the noise in the individual bird-brain navigation systems would cancel out. The emergent entity – the flock – in turn would affect the individual geese, helping them to navigate better than they could on their own.

When people pray, they ask for intervention on their or others' behalf. The view that I've been outlining invites us to think of the effects of prayer as akin to becoming more sensitive to the need to stay close to the other birds in the flock: By praying, one can become more sensitive to the emergent supreme being. Such increased sensitivity may imply that one can contribute more strongly to this emergent entity.

By analogy, it's as if one of those geese became aware of the 'keep near' rule and decided to nudge the other birds in a particular direction – which thereby allows it to influence the flock's effect on itself. To the extent that prayer puts one closer to God, one's plea for intervention will have a larger impact on the way that the Ultimate Superset exerts downward causality. But note that according to this view God works rather slowly. Think of dropping rocks in a pond: It takes time for the ripples to propagate and be reflected back from the edge, forming interference patterns in the center of the pond.

A crucial idea in monotheistic religions is that God is the Creator. The present approach may help us begin to grapple with this idea, as follows:

First, consider each individual person. The environment plays a key role in creating who and what we are, because there are far too few genes to program every aspect of our brains. For example, when you were born, your genes programmed many connections in your visual areas but did not specify the precise circuits necessary to determine how far away objects are. As an infant, the act of reaching for an object tuned the brain circuits that estimate how far away the object was from you. Similarly, your genes enabled you to acquire language but not a specific language. The act of acquiring a language shapes your brain – which in turn may make it difficult to acquire another language, with different sounds and grammar, later in life. Moreover, cultural practices configure the brains of members of the culture. (A case in point: The Japanese have many forms of bowing – forms that are difficult for a Westerner to master relatively late in life. When we try to bow, we bow with an accent.) And the environment plays not only an essential role in how we develop as children but also a continuing role in how we develop over the course of our lives. The act of learning literally changes who and what we are.

According to this perspective, it's not just negotiating the physical world and sociocultural experience that shape the brain: The Ultimate Superset – the emergent property of all living things – affects all of the influences that make us who and what we are, both as we develop during childhood and continue to learn and develop as adults.

Next, consider our species. One could try to push this perspective into a historical context and note that evolution by natural selection reflects the effects of interactions among living things. If so, then the emergent properties of such interactions could feed back to affect the course of evolution itself.

In short, it is possible to begin to view the divine through the lens of science. But such reasoning does no more than set the stage. To be a truly dangerous idea, this sort of proposal must be buttressed by the results of empirical test. At present, my point is not to convince but rather to intrigue. As much as I admired Stephen Jay Gould (and I did, very much), perhaps he missed the mark on this one. Perhaps there is a grand project waiting to be launched, to integrate the two great sources of knowledge and belief in the world today – science and religion.

Science Will Never Silence God

JESSE BERING

Jesse Bering is director of the Institute of Cognition and Culture at the Queen's University, Belfast.

With each meticulous turn of the screw in science, with each tightening up of our understanding of the natural world, we pull more taut the straps over God's muzzle. From botany to bioengineering, from physics to psychology, what is science, really, but true revelation – and what is revelation but the negation of God? It is a humble pursuit we scientists engage in: racing toward reality. Many of us suffer the harsh glare of the American theocracy, whose heart still beats loud and strong in this new era of the twenty-first century. We bravely favor truth, in all its wondrous, amoral, and meaningless complexity, over the singularly destructive Truth born out of the trembling minds of our ancestors. But my dangerous idea, I fear, is that no matter how far our thoughts vault into the eternal sky of scientific progress, no matter how dazzling the effects of this progress, God will always bite through his muzzle and banish us from the starry night of humanistic ideals.

Science is an endless series of binding and rebinding his breath. There will never be a day when God does not speak for the majority. There will never even be a day when he does not whisper into the ears of the most godless of scientists. This is because God is not an idea, nor a cultural invention, nor an 'opiate of the masses,' nor any such thing. God is *a way of thinking* that has been rendered permanent by natural selection.

As scientists, we must toil and labor and toil again to silence God, but ultimately this is like cutting off our ears to hear more clearly. God, too, is a biological appendage. Until we acknowledge this fact, until we rear our children with this knowledge, he will continue to howl his discontent for all of time.

Religion Is the Hope That Is Missing in Science

SCOTT ATRAN

Scott Atran is a research director in anthropology at the CNRS in Paris, visiting Professor of Psychology and Public Policy at the University of Michigan, Presidential Scholar at the John Jay School of Criminal Justice, and the author of *In Gods We Trust: The Evolutionary Landscape of Religion*.

Religion, like mathematics, is an evolutionary by-product of various mental faculties of the human brain that most people, in all known societies, intermittently converge on with differing degrees of intensity as they interact with the world. Whereas mathematics describes fundamental interactions with (non-intentional) objects (including objects of thought), religion manages fundamental interactions with (intentional) subjects, by establishing the moral foundations for existence, death, and key segments in the intervening life cycle. Like mathematics, it can be used and studied in many different ways, including from the vantage point of cognitive science. But the fact that it can be objectively studied by no means implies that its subjective importance to human life is any less.

I find it fascinating that brilliant scientists and philosophers have no clue about how to deal with the basic irrationality of human life and society other than to insist, against all reason and evidence, that things ought to be rational and evidence-based. Makes me embarrassed to be an atheist.

I find no historical evidence whatever that scientists have a keener or deeper appreciation than religious people of how to deal with personal or moral problems. Some scientists have some good and helpful insights into existential problems some of the time, but some good scientists have done more to harm others than most people are remotely capable of.

True, some people operating in the name of religion have been more explicitly savage and cruel toward others than most, but there are the likes of Lincoln, Gandhi and Martin Luther King, whose religion not only has given hope to so many but has thereby cumulatively enabled the lessening of human misery.

Ever since Edward Gibbon's *Decline and Fall of the Roman Empire*, scientists and secularly minded scholars have been predicting the ultimate demise of religion. But if anything, religious fervor is increasing across the world, including in the United States, the world's most economically powerful and scientifically advanced society. An underlying reason is that science treats humans and intentions only as incidental elements in the universe, whereas for religion they are central. Science is not particularly well suited to deal with people's existential anxieties, including death, deception, sudden catastrophe, loneliness, or longing for love or justice. It cannot tell us what we ought to do, only what we can do. Religion thrives because it addresses people's deepest emotional yearnings and society's foundational moral needs, perhaps even more so in complex and mobile societies that are increasingly divorced from nurturing family settings and long familiar environments.

From a scientific perspective of the overall structure and design of the physical universe:

(1) Human beings are accidental and incidental products of the material development of the universe, almost wholly irrelevant and readily ignored in any general description of its functioning.

   Beyond Earth there is no intelligence – however alien or like our own – that is watching out for us or cares. We are alone.

(2) Human intelligence and reason, which searches for the hidden traps and causes in our surroundings, evolved and will always remain leashed to our animal passions in the struggle for survival, the quest for love, the yearning for social standing and belonging. This intelligence does not easily tolerate loneliness, any more than it tolerates the looming prospect of death (individual or collective).

Religion is the hope that is missing in science.

But doesn't religion impede science, and vice versa? Not necessarily. Leaving aside the sociopolitical stakes in the opposition between science and religion (which vary widely and are not constitutive of science or religion per se; Calvin considered obedience to tyrants as exhibiting trust in God, Franklin wanted the motto of the American Republic to be 'Rebellion against tyranny is obedience to God'), a crucial difference between science and religion is that factual knowledge as such is not a principal aim of religious devotion but plays merely a supporting role. Only in the last decade has the Catholic Church reluctantly acknowledged the factual plausibility of Copernicus, Galileo, and Darwin. Earlier religious rejection of their theories stemmed from challenges posed to a cosmic order unifying the moral and material worlds. Separating out the core of the material world would be like draining the pond where a water lily grows. A long lag time was necessary to refurbish and remake the moral and material connections in such a way that would permit faith in a unified cosmology to survive.

Myths and Fairy Tales Are Not True

TODD E. FEINBERG

Todd E. Feinberg is a psychiatrist and neurologist at the Albert Einstein College of Medicine and the author of *The Lost Self: Pathologies of the Brain and Identity*.

'Myths and fairy tales are not true.' There is no Easter Bunny, there is no Santa Claus, and Moses may never have existed. Worse yet, I have increasing difficulty believing that there is a higher power ruling the universe. This is my dangerous idea. It is not a dangerous idea to those who do not share my particular worldview or personal fears; to others it may seem trivially true. But for me, this idea is downright horrifying.

I came to ponder this idea through my neurological examination of patients with brain damage that causes a disturbance in their self-concepts and ego functions.

Some of these patients develop, in the course of their illness and recovery (or otherwise), disturbances of self- and personal relatedness that create enduring delusions and metaphorical confabulations regarding their bodies, their relationships with loved ones, and their personal experiences. A patient I examined with a right hemisphere stroke and paralyzed left arm claimed that the arm was actually severed from his brother's body by gang members, thrown in the East River, and later attached to the patient's shoulder. Another patient with a ruptured brain aneurysm and amnesia who denied his disabilities claimed he was planning to adopt a (phantom) child who was in need of medical assistance.

These personal narratives, produced by patients in altered neurological states and therefore without the constraints imposed by a fully functioning consciousness, have a dreamlike quality and constitute personal myths that express the patients' beliefs about themselves. The patient creates a metaphor in which personal experiences are crystallized in the form of external real or fictitious persons, objects, places, or events. When this occurs, the metaphor serves as a symbolic representation or externalization of the patient's feelings that the patient does not realize originates from within the self.

There is an intimate relationship between my patients' narratives and socially endorsed fairy tales and mythologies. This is particularly apparent when the mythologies deal with themes relating to a loss of self or personal identity, or to death. For many people, the notion of personal death is extremely difficult to grasp and fully accommodate within the self-image. For many, in order to go on with life, death must be denied. Therefore, to help the individual deal with the prospect of the inevitability of personal death, cultural and religious institutions provide metaphors of everlasting life. Just as my patients adapt to difficult realities by creating metaphorical substitutes, it appears to me that beliefs in angels, deities, and eternal souls can be understood in part as wish-fulfilling metaphors for an unpleasant reality that most of us cannot fully comprehend and accept.

Parental Licensure

DAVID LYKKEN

David Lykken is a behavioral geneticist and emeritus professor of psychology at the University of Minnesota. He is the author of *Happiness*.

I believe that during my grandchildren's lifetimes the U.S. Supreme Court will find a way to approve laws requiring parental licensure.

Traditional societies in which children are socialized collectively, the method to which our species is evolutionarily adapted, have very little crime. In the modern United States, the proportion of fatherless children, living with unmarried mothers, currently some 10 million, has increased more than 400 percent since 1960, while the violent crime rate had risen to 500 percent by 1994 before dipping slightly (due to an increase in the number of prison inmates). In 1990, across the fifty states, the correlation between the violent crime rate and the proportion of illegitimate births was 0.70.

About 70 percent of incarcerated delinquents, teenage pregnancies, adolescent runaways involve (I think result from) fatherless rearing. Because these frightening curves continue to accelerate, I believe we must eventually confront the need for parental licensure – you can't keep that newborn unless you are twenty-one, married, and self-supporting – not just for society's safety but so those babies will have a chance for life, liberty, and the pursuit of happiness.

Zero Parental Influence

JUDITH RICH HARRIS

Judith Rich Harris is an independent investigator and theoretician. She is the author of *The Nurture Assumption* and *No Two Alike: Human Nature and Human Individuality*.

Is it dangerous to claim that parents have no power at all (other than genetic) to shape their child's personality, intelligence, or the way he or she behaves outside the family home? More to the point, is this claim false? Was I wrong when I proposed that parents' power to do these things by environmental means is zero, *nada*, zilch?

A confession: When I first made this proposal ten years ago, I didn't fully believe it myself. I took an extreme position – the null hypothesis of zero parental influence – for the sake of scientific clarity. Making myself an easy target, I invited the establishment – research psychologists in the academic world – to shoot me down. I didn't think it would be all that difficult for them to do so. It was clear by then that there weren't any big effects of parenting, but I thought there must be modest effects that I would ultimately have to acknowledge.

The establishment's failure to shoot me down has been nothing short of astonishing. One developmental psychologist even admitted recently that researchers hadn't yet found proof that 'parents do shape their children' but she was still convinced they would eventually find it if they kept searching long enough. Her comrades-in-arms have been less forthright. 'There are dozens of studies that show the influence of parents on children!' they kept saying, but then they would neglect to name them, perhaps because these studies were among ones I had already demolished by showing that they lacked the necessary controls or the proper statistical analyses. Or they would claim to have newer research that provided an airtight case for parental influence, but again there was a catch: The work had never been published in a peer-reviewed journal. When I investigated, I could find no evidence that the research in question had actually been done – or, if done, that it had produced the results claimed for it. At most, it appeared to consist of preliminary work with too little data to be meaningful (or publishable).

Vaporware, I call it. Some of the vaporware has achieved mythic status. You may have heard of Stephen Suomi's experiment with nervous baby monkeys, supposedly showing that those reared by 'nurturant' adoptive monkey mothers turn into calm, socially confident adults. Or of Jerome Kagan's research with nervous baby humans, supposedly showing that those reared by 'overprotective' (that is, nurturant) human mothers are more likely to remain fearful.

Researchers like these might well see my ideas as dangerous. But is the notion of zero parental influence dangerous in any other sense? So it is alleged. Here's what Frank Farley, former president of the American Psychological Association, told a journalist in 1998: '[Harris's] thesis is absurd on its face, but consider what might happen if parents believe this stuff! Will it free some to mistreat their kids, since "it doesn't matter"? Will it tell parents who are tired after a long day that they needn't bother even paying any attention to their kid since "it doesn't matter"?' Farley seems to be saying that the only reason parents are nice to their children is because they think it will make the children turn out better! And that if parents believed that they had no influence at all on how their kids turn out, they are likely to abuse or neglect them.

Which, it seems to me, is absurd on its face. Most chimpanzee mothers are nice to their babies and take good care of them. Do chimpanzees think they're going to influence how their offspring turn out? Doesn't Frank Farley know anything at all about evolutionary biology and evolutionary psychology?

My idea is viewed as dangerous by the powers that be, but I don't think it's dangerous at all. On the contrary: If people accepted it, it would be a breath of fresh air. Family life, for parents and children alike, would improve. Look what's happening now as a result of the faith (obligatory in our culture) in the power of parents to mold their children's fragile psyches. Parents are exhausting themselves in their efforts to meet their children's every demand, not realizing that evolution designed offspring – nonhuman animals as well as humans – to demand more than they really need. Family life has become phony, because parents are convinced that children need constant assurances of parental love, so if they don't happen to feel very loving at a particular time or toward a particular child, they fake it. Praise is delivered by the bushel, which devalues its worth. Children have become the masters of the home.

And what has all this sacrifice and effort on the part of parents bought them? Zilch! There are no indications that children today are happier, more self-confident, less aggressive, or in better mental health than they were sixty years ago, when I was a child – when homes were run by and for adults, when physical punishment was used routinely, when fathers were generally unavailable, when praise was a rare and precious commodity, and when explicit expressions of parental love were reserved for the deathbed.

Is my idea dangerous? I've never condoned child abuse or neglect; I've never believed that parents don't matter. The relationship between a parent and a child is an important one, but it's important in the same way as the relationship between married partners. A good relationship is one in which each party cares about the other and derives happiness from making the other happy. A good relationship is not one in which one party's central goal is to modify the other's personality.

What's really dangerous – perhaps a better word is tragic – is the establishment's idea of the all-powerful, and hence all-blamable, parent.

The Focus on Emotional Intelligence

JOHN GOTTMAN

John Gottman is a psychologist and the founder of the Gottman Institute, in Seattle, Washington. He is the author, most recently, of *The Relationship Cure, A 5-Step Guide for Building Better Connections with Family, Friends, and Lovers.*

The most dangerous idea I know of is emotional intelligence. Within the context of the cognitive neuroscience revolution in psychology, the focus on emotions is extraordinary. The overarching idea that there is such a thing as emotional intelligence, that it has a neuroscience, that it is inter-personal, i.e. between two brains, rather than within one brain, are all quite revolutionary concepts about human psychology. It is also a revolution in thinking about infancy, couples, family, adult development, aging, and so on.

A Cacophony of 'Controversy'

ALISON GOPNIK

Alison Gopnik is a psychologist at the University of California at Berkeley and coauthor (with Andrew N. Meltzoff and Patricia K. Kuhl) of *The Scientist in the Crib: What Early Learning Tells Us About the Mind*.

It may not be a good idea to encourage scientists to articulate dangerous ideas.

Good scientists, almost by definition, tend toward the contrarian and ornery; nothing gives them more pleasure than holding to an unconventional idea in the face of opposition. Indeed, orneriness and contrarianism are something of currency for science – nobody wants to have an idea that everyone else has, too. Scientists are always constructing a straw-man establishment opponent whom they can then fearlessly demolish. If you combine that with defying the conventional wisdom of nonscientists, you have a recipe for a distinctive kind of scientific smugness and self-righteousness. We scientists see this contrarian habit grinning back at us in a particularly hideous and distorted form when global-warming opponents or intelligent-design advocates invoke the unpopularity of their ideas as evidence that they should be accepted, or at least discussed.

The problem is exacerbated for public intellectuals. For the media, too, would far rather hear about contrarian or unpopular or morally dubious or 'controversial' ideas than about ideas that are congruent with everyday morality and wisdom. No one writes a newspaper article about a study showing that girls are just as good at some task as boys, that there aren't IQ differences between races, or that children are influenced by their parents.

It is certainly true that there is no reason that scientifically valid results should have morally comforting consequences — but there's no reason why they shouldn't, either. Unpopularity or shock is no more a sign of truth than popularity is. More to the point, when scientists do have ideas that are potentially morally dangerous, they should approach those ideas with hesitancy and humility. And they should do so in full recognition of the great human tragedy that, as Isaiah Berlin pointed out, there can be genuinely conflicting goods and that humans are often in situations of conflict for which there is no simple or obvious answer.

Truth and morality may indeed in some cases be competing values, but that is a tragedy, not a cause for self-congratulation. Humility and empathy come less easily to most scientists (most certainly including me) than pride and self-confidence, but perhaps for that very reason those are the virtues we should pursue.

This is, of course, itself a dangerous idea. Orneriness and contrarianism are genuine scientific virtues, too. And in the current profoundly antiscientific political climate, it is terribly dangerous to do anything that might give comfort to the enemies of science. But I think the peril to science actually doesn't lie in timidity or self-censorship. It is much more likely to lie in a cacophony of 'controversy.'

Applied History

STEWART BRAND

Stewart Brand is the founder of the *Whole Earth Catalog* and the author of *The Clock of the Long Now*.

All historians understand that they must never, ever talk about the future. Their discipline requires that they deal in facts, and the future doesn't have any yet. A solid theory of history might be able to embrace the future, but all such theories have been discredited. Thus historians do not offer, and are seldom invited, to take part in shaping public policy. They leave that to economists.

But discussions among policy makers always invoke history anyway, usually in simplistic form. 'Munich' and 'Vietnam,' devoid of detail or nuance, stand for certain kinds of failure. 'Marshall Plan' and 'Men on the Moon' stand for certain kinds of success. Such totemic invocation of history is the opposite of learning from history, and Santayana's warning continues in force: that those who fail to learn from history are condemned to repeat it.

A dangerous thought: What if public policy makers have an obligation to engage historians, and historians have an obligation to try to help?

And instead of just retailing advice, go generic. Historians could set about developing a rigorous subdiscipline called applied history.

There is only one significant book on the subject, published in 1988. *Thinking in Time: The Uses of History for Decision Makers* was written by the late Richard Neustadt and Ernest May, who long taught a course on the subject at Harvard's Kennedy School of Government. (A course called 'Reasoning from History' is currently taught there by Alexander Keyssar.)

Done wrong, applied history could paralyze public decision making and corrupt the practice of history – that's the danger. But done right, applied history could make decision making and policy far more sophisticated and adaptive, and it could invest the study of history with the level of consequence it deserves.

Tribal Peoples Often Damage Their Environments and Make War

JARED DIAMOND

Jared Diamond is a biologist and a geographer at UCLA. His latest book is *Collapse: How Societies Choose to Fail or Succeed*.

Why is this idea dangerous? Because too many people today believe that a reason not to mistreat tribal peoples is that they are too nice or wise or peaceful to do those evil things, which only we evil citizens of state governments do. The idea is dangerous because, if you believe that that's the reason not to mistreat tribal peoples, then proof of the idea's truth would suggest that it's OK to mistreat them. In fact, the evidence seems to me overwhelming that the dangerous idea is true. But we should treat other people well because of ethical reasons, not because of naïve anthropological theories that will almost surely prove false.

Nothing

CHARLES SEIFE

Charles Seife is a professor of journalism at New York University and the author of *Zero: The Biography of a Dangerous Idea*.

Nothing can be more dangerous than nothing.

Humanity has always been uncomfortable with zero and the void. The ancient Greeks declared them unnatural and unreal. Theologians argued that God's first act was to banish the void by the act of creating the universe *ex nihilo*, and medieval thinkers tried to ban zero and the other Arabic 'ciphers.' But the emptiness is all around us. Most of the universe is void. Even as we huddle around our hearths and invent stories to convince ourselves that the cosmos is warm and full and inviting, nothingness stares back at us with empty eye sockets.

Everything Is Pointless

SUSAN BLACKMORE

Susan Blackmore is a psychologist, a skeptic, and the author of, among other books, *Consciousness: An Introduction*.

We humans can and do make up our own purposes, but ultimately the universe has none. All the wonderfully complex and beautifully designed things we see around us were built by the same purposeless process: evolution by natural selection. This includes everything from microbes and elephants to skyscrapers and computers and even our own inner selves.

People have mostly got used to the idea that living things were designed by natural selection, but they have more trouble accepting the idea that human creativity is exactly the same process operating on memes – the units of cultural information – instead of genes. It seems, they think, to take away uniqueness, individuality, and true creativity.

Of course it does nothing of the kind; each person is unique, even if that uniqueness is explained by that individual's particular combination of genes, memes, and environment rather than by an inner conscious self who is the fount of creativity. So I think it is true (but is it dangerous?) to say this: You may think that I wrote this piece, but in fact it was written by memes competing in the pointless universe.

There Aren't Enough Minds to House the Population Explosion of Memes

DANIEL C. DENNETT

Daniel C. Dennett is a philosopher, university professor, and codirector of the Center for Cognitive Studies at Tufts University. He is the author, most recently, of *Breaking the Spell: Religion As a Natural Phenomenon*.

Ideas can be dangerous. Darwin had one, for instance. We hold all sorts of inventors and other innovators responsible for assaying, in advance, the environmental impact of their creations, and since ideas can have huge environmental impacts, I see no reason to exempt us thinkers from the responsibility of quarantining any deadly ideas we may happen to come across.

So if I found what I took to be such a dangerous idea, I would button my lip until I could find some way of preparing the ground for its safe expression. I expect that others who are replying to this year's *Edge* question have engaged in similar reflections and arrived at the same policy. If so, then some people may be pulling their punches with their replies. The really dangerous ideas they are keeping to themselves.

But here is an unsettling idea that is bound to be true in one version or another, and as far as I can see, it won't hurt to publicize it. It might well help.

The human population is still growing but at nowhere near the rate that the population of memes is growing. There is competition for the limited space in human brains for memes, and something has to give. Thanks to our incessant and often technically brilliant efforts, and our apparently insatiable appetites for novelty, we have created an explosively growing flood of information, in all media, on all topics, in every genre. Now, either (1) we will drown in this flood of information, or (2) we won't drown in it. Both alternatives are deeply disturbing. What do I mean by drowning? I mean that we will become psychologically overwhelmed, unable to cope, victimized by the glut and unable to make life-enhancing decisions in the face of an unimaginable surfeit. (I recall the brilliant scene in the film of Evelyn Waugh's dark comedy *The Loved One* in which embalmer Mr. Joyboy's gluttonous mother is found sprawled on the kitchen floor, helplessly wallowing in the bounty that has spilled from a capsized refrigerator.) We will be lost in the maze, preyed upon by whatever clever forces find ways of pumping money – or simply further memetic replications – out of our situation. In *The War of the Worlds*, H. G. Wells sees that it might well be our germs, not our high-tech military contraptions, that subdue our alien invaders. Similarly, might our own minds succumb not to the devious manipulations of evil brainwashers and propagandists but to nothing more than a swarm of irresistible ditties, *Nous* nibbled to death by slogans and one-liners?

If we don't drown, how will we cope? If we somehow learn to swim in the rising tide of the infosphere, that will entail that we – that is, our grandchildren and their grandchildren – become very very different from our recent ancestors. What will 'we' be like? (Some years ago, Douglas Hofstadter wrote a wonderful piece, 'In 2093, Just Who Will Be We?' in which he imagines robots being created to have 'human' values – robots that gradually take over the social roles of our biological descendants, who have become stupider and less concerned with the things *we* value. If we could secure the welfare of just one of these groups, our children or our brainchildren, which group would we care about the most? With which group would we identify?)

Whether 'we' are mammals or robots in the not so distant future, what will we know and what will we have forgotten forever, as our previously shared intentional objects recede in the churning wake of the great ship that floats on this sea and charges into the future propelled by jets of newly packaged information? What will happen to our cultural landmarks? Presumably our descendants will all still recognize a few reference points (the pyramids of Egypt, arithmetic, the Bible, Paris, Shakespeare, Einstein, Bach . . .) but as wave after wave of novelty passes over them, what will they lose sight of? The Beatles are truly wonderful, but if their cultural immortality is to be purchased by the loss of such minor twentieth-century figures as Billie Holiday, Igor Stravinsky, and Georges Brassens, what will remain of our shared understanding?

The intergenerational mismatches we all experience in macroscopic versions (Great-Grandpa's joke falls on deaf ears, because nobody else in the room knows that Nixon's wife was named Pat) will presumably be multiplied to the point where much of the raw information we have piled in our digital storehouses is simply incomprehensible to everyone – except that we will have created phalanxes of 'smart' Rosetta stones of one sort or another that can 'translate' the alien material into something we (think maybe we) understand. I suspect we hugely underestimate the importance (to our sense of cognitive security) of our regular participation in the four-dimensional human fabric of mutual understanding, with its reassuring moments of shared – and *seen* to be shared, and *seen* to be *seen* to be shared – comprehension.

What will happen to common knowledge in the future? I do think our ancestors had it easy: Aside from all the juicy bits of unshared gossip and some proprietary trade secrets and the like, people all knew pretty much the same things and knew that they knew the same things. There just wasn't that much to know. Won't people be able to create and exploit *illusions* of common knowledge in the future, virtual worlds in which people only think they are in touch with their cyberneighbors?

I see small-scale projects that might protect us to some degree, if they are done wisely. Think of all the work published in academic journals before, say, 1990 that is in danger of becoming practically invisible to later researchers because it can't be found on-line with a good search engine. Just scanning it all and hence making it 'available' is not the solution. There is too much of it. But we could start projects in which (virtual) communities of retired researchers who still have their wits about them and who know particular literatures well could brainstorm among themselves, using their pooled experience to elevate the forgotten gems, rendering them accessible to the next generation of researchers. This sort of activity has in the past been seen to be a stodgy sort of scholarship, fine for classicists and historians but not fit work for cutting-edge scientists and the like. I think we should try to shift this imagery and help people recognize the importance of providing for one another this sort of pathfinding through the forests of information. It's a drop in the bucket, but perhaps if we all start thinking about conservation of valuable mind-space, we can save ourselves (our descendants) from informational collapse.

Unspeakable Ideas

RANDOLPH M. NESSE

Randolph M. Nesse is a professor of psychiatry at the University of Michigan. His latest book (coedited with Deborah Carr and Irene Wortman) is *Spousal Bereavement in Late Life*.

The idea of promoting dangerous ideas seems dangerous to me. I spend considerable effort to prevent my ideas from becoming dangerous – except, that is, to entrenched false beliefs and to myself. For instance, my idea that bad feelings are useful for our genes upends much conventional wisdom about depression and anxiety. I find, however, that I must firmly restrain journalists who are eager to share the sensational but incorrect conclusion that depression should not be treated. Similarly, many people draw dangerous inferences from my work on Darwinian medicine. For example, just because fever is useful does not mean that it should not be treated. I now emphasize that evolutionary theory does not tell you what to do in the clinic, it just tells you what studies need to be done.

I also feel obligated to prevent my ideas from becoming dangerous on a larger scale. For instance, many people who hear about Darwinian medicine assume incorrectly that it implies support for eugenics. I encourage them to read history as well as my writings. The record shows how quickly natural selection was perverted into social Darwinism, an ideology that seemed to justify letting poor people starve. Related ideas keep emerging. We scientists have a responsibility to challenge dangerous social policies incorrectly derived from evolutionary theory. Racial superiority is yet another dangerous idea that hurts real people. More examples come to mind all too easily, and some quickly get complicated. For instance, the idea that men are inherently different from women has been used to justify discrimination, but the idea that men and women have identical abilities and preferences may also cause great harm.

While I don't want to promote ideas that are dangerous to others, I am fascinated by ideas that are dangerous to anyone who expresses them. These are 'unspeakable ideas.' By unspeakable ideas, I don't mean those whose expression is forbidden in a certain group. Instead, I propose that there is class of ideas whose expression is inherently dangerous everywhere and always, because of the nature of human social groups. Such unspeakable ideas are antimemes. Memes, both true and false, spread fast because they are interesting and give social credit to those who spread them. Unspeakable ideas – even true, important ones – don't spread at all, because expressing them is dangerous to those who do so.

So why, you may ask, is a sensible scientist even bringing the idea of unspeakable ideas up? Isn't the idea of unspeakable ideas a dangerous idea? I expect I will find out. My hope is that a thoughtful exploration of unspeakable ideas will not hurt people in general, perhaps won't hurt me much, and might unearth some long neglected truths.

Generalizations cannot substitute for examples, even if providing examples is risky. So, please gather your own data. Here is an experiment. The next time you are having a drink with an enthusiastic fan for your hometown team, say, 'Well, I think our team just isn't very good and didn't deserve to win.' Or, moving to more risky territory, when your business group is trying to deal with a savvy competitor, say, 'It seems to me that their product is superior, because they are smarter than we are.' Finally – and I cannot recommend this, but it offers dramatic data – you could respond to your spouse's difficulties at work by saying, 'If they are complaining about you not doing enough, it is probably because you just aren't doing your fair share.' Most people do not need to conduct such social experiments to know what happens when such unspeakable ideas are spoken.

Many broader truths are equally unspeakable. Consider, for instance, all the articles written about leadership. Most are infused with admiration and respect for a leader's greatness. Much rarer are articles about the tendency for leadership positions to be attained by power-hungry men who use their influence to further advance their self-interest. Then there are all the writings about sex and marriage. Most of them suggest that there is some solution that allows full satisfaction for both partners while maintaining secure relationships. Questioning such notions is dangerous, unless you are a comic, in which case skepticism can be very, very funny.

As a final example, consider the unspeakable idea of unbridled self-interest. Someone who says, 'I will only do what benefits me,' has committed social suicide. Tendencies to say such things have been selected against, while those who advocate goodness, honesty, and service to others get wide recognition. This creates an illusion of a moral society, which then, thanks to the combined forces of natural and social selection, becomes a reality that makes social life vastly more agreeable.

There are many more examples, but I must stop here. To say more would either get me in trouble or falsify my argument. Will I ever publish my *Unspeakable Essays*? It would be risky, wouldn't it?

Anty Gravity: Chaos Theory in an All-Too-Practical Sense

KAI KRAUSE

Kai Krause is a philosopher, artist, and software developer and the author of *3D Science: New Scanning Electron Microscope Imagery*.

Dangerous ideas? It is dangerous ideas you want? From this group of people? That in itself ought to be nominated as one of the more dangerous ideas.

Danger is ubiquitous. If recent years have shown us anything, it should be that very simple small events can cause havoc in our society. A few hooded youths play cat-and-mouse with the police: *Bang!*, thousands of burned cars put all of Paris into a state of paralysis and mandatory curfew and the entire system into a state of shock and horror.

My first thought was: What if any *really* smart set of people set their mind to it . . . How utterly and scarily trivial it would be to disrupt the very fabric of life, to bring society to a dead stop!

The relative innocence and stability of the last fifty years may spiral into a nearly inevitable exposure to real chaos. What if it isn't haphazard testosterone-driven riots, where the rioters cannibalize their own neighborhood, much as in Los Angeles in the 1980s, but someone with real insight behind that criminal energy? What if slashdotters start musing aloud, 'Gee, the L.A. water supply is rather simplistic, isn't it?' An Open Source crime web, a Wiki for real WTO opposition? Hacking L.A. may be a lot easier than hacking Internet Explorer.

That's basic banter over a beer in a bar. I don't even want to speculate about what a serious set of brainiacs could conjure up, and I refuse to give it any more print space here. However, the danger of such sad memes is what requires our attention.

In fact, I will broaden the specter still more: It's not violent crime and global terrorism I worry about as much as the underpinning of our entire civilization coming apart. No acts of malevolence, no horrible plans by evil dark forces, neither the singular 'Bond Nemesis' kind nor masses of religious fanatics. None of that is needed. It is the *glue* that is coming apart to topple this tower. And no, I am not referring to 'spiraling trillions of debt.'

No, what I am referring to is a slow process I have observed over the last thirty years, ever since (in my teens) I began to wonder, 'How would this world work if everyone were like me?' and realized that it wouldn't! It was amazing to me that there were just enough people to make just enough shoes so that everyone could avoid walking barefoot. That there were people volunteering to spend their time day-in-day-out being dentists and lawyers and salesmen. For almost *any* 'job' job I look at, I have the sincerest admiration for the tenacity of the people who do them . . . How *do* they do it? It would drive me nuts after *hours*, let alone years . . . Who makes those shoes? Who drills those teeth?

That was my wondrous introspection in my adolescent phase, while I was searching for a place in the jigsaw puzzle.

But in recent years, the haunting question has come back to me: 'How the hell *does* this world function at all? And does it, really?' I feel an alienation zapping through the channels; I can't find myself connecting with those groups of humanoids trouncing around MTV. Especially the glimpses of 'real life' on daytime courtroom dramas – or just looking at faces in the street – on every scale, the more closely I observe, the more the creeping realization haunts me: Individuals, families, groups, neighborhoods, cities, states, countries all just barely hang in there between debt and dysfunction. The whole planet looks like Anytown, with minimalls cutting up the landscape, and just down the road it's all white trash with rusty car wrecks in the backyards. A huge Groucho club I don't want to be a member of.

But it does go further: What is particularly disturbing to see is this desperate search for what I call Super-Individualism, which has rampantly increased in the last decade or so.

Everyone suddenly needs to be *so* special, be utterly unique. So unique that they race off like lemmings to get ever more 'individual' tattoos – branded cattle, with branded chains in every mall, converging on a bland sameness worldwide. Every rap singer with ever more gold chains in ever longer stretch limos is singing the tune: 'Don't be a loser! Don't be normal!'

But now the tables are turning: the anthill is relying on the behavior of the ants to function properly. And that requires social behavior, role playing, taking defined tasks and following them through.

What if each ant suddenly wants to be the queen? What if soldiering and nest building and cleaning chores is just not cool enough any more?

If AntTV shows them, every day, nothing but *un*-Ant behavior . . .?

In my youth we whined about what to do and how to do it, but in the end all my friends became 'normal' humans – orthopedic surgeons and professors, social workers, designers . . . There were always a few who lived on the edges of normality, ending up as television celebrities, but on the whole they were perfectly reasonable ants: 1.8 children, 2.7 cars, 3.3 TVs . . .

Now I am no longer confident that that line can continue. If every honeymoon is now booked in Bali on a Visa card and every kid in Borneo wants to play ball in NYC, can the network of society be pliable enough to accommodate total upheaval? And what if 2 billion Chinese and Indians raise a generation of kids staring 6+ hours a day into All-American values they can never attain . . . taunted by Hollywood movies about heroic acts and pathetic dysfunctionality coupled with ever increasing violence and disdain for ethics or morals.

Seeing scenes of desperate youths in South American slums watching *Kill Bill* makes me think, 'This is just oxygen thrown into the fire!' The ants will not play along much longer. The anthill will not survive if even a small fraction of the system is falling apart.

Couple that inane drive for Super Individualism (and the Quest for Coolness by an ever increasing group destined to fail miserably) with the scarily simple realization of how effective even a small set of desperate people can become, then add the obvious penchant for religious fanaticism, and you have an ugly picture of the long-term future.

So many curves that grow upward toward limits, so many statistics that show increases and no way to turn around.

Many contributors to this forum may speculate about infinite life spans, changing the speed of light, finding ways to decode consciousness, wormholes to other dimensions, and grand unified theories. I applaud that! It does take all kinds. Those are viable and necessary questions for humankind as a whole.

However, I believe we need to clean house, re-evaluate, re-define the priorities. While we look at the horizon here in these pages, it is the ground beneath us that may be crumbling. The anthill could go to ant Hell!

Next year, let's ask for *good* ideas, not dangerous ones. Really practical, serious, *good ideas,* like: 'What is the most immediate positive global impact of any kind that can be achieved within one year?' 'How do we envision Internet3 and Web3 as a real platform for a global brainstorming with 6+ billion potential participants.'

This was not meant to sound like doom-and-gloom naysaying. I see myself as a sincere optimist but one who believes in realistic pessimism as a useful tool to initiate change.

Navigating by New Scientific Principles

RUPERT SHELDRAKE

Rupert Sheldrake is a biologist living in London. His latest book is *The Sense of Being Stared At: And Other Aspects of the Extended Mind.*

We don't understand animal navigation.

No one knows how pigeons home, or how swallow migrate, or how green turtles find Ascension Island from thousands of miles away to lay their eggs. These kinds of navigation involve more than following familiar landmarks or orienting oneself in a particular compass direction; they involve an ability to move toward a goal.

Why is this current ignorance of ours dangerous? Don't we just need a bit more time to explain that navigation in terms of standard physics, genes, nerve impulses, brain chemistry? Perhaps.

But there is a possibility that animal navigation may not be explicable in terms of present-day physics. Over and above the known senses, some species of animals may have a sense of direction that depends on their being attracted toward their goals through direct, fieldlike connections. These spatial attractors are places with which the animals themselves are already familiar or with which their ancestors were familiar.

What are the facts? We know more about pigeons than any other species. Within familiar territory, especially within a few miles of their home, pigeons can use landmarks; for example, they can follow roads. But using familiar landmarks near home cannot explain how racing pigeons return across unfamiliar terrain from six hundred miles away, even flying over the sea, as English pigeons do when they are raced from Spain.

Charles Darwin, himself a pigeon fancier, was one of the first to suggest a scientific hypothesis for pigeon homing. He proposed that they might use a kind of dead reckoning, registering all the twists and turns of the outward journey. This idea was tested many years later, by taking pigeons away from their lofts in closed vans by devious routes. They still homed normally. So did birds transported on rotating turntables. So did birds that had been completely anaesthetized during the outward journey.

What about celestial navigation? One problem for hypothetical solar or stellar navigation systems is that many animals continue to navigate in cloudy weather. Another problem is that celestial navigation depends on a precise time sense. To test the sun navigation theory, homing pigeons were clock-shifted by six or twelve hours and taken many miles from their lofts before being released. On sunny days, they set off in the wrong direction, as if a clock-dependent sun compass had been shifted. But they soon corrected their course and flew homeward normally.

Two main hypotheses remain: smell and magnetism. Smelling the home position from hundreds of miles away is generally agreed to be implausible. Even the most ardent defenders of the smell hypothesis (the Italian school of Floriano Papi and his colleagues) concede that smell navigation is unlikely to work at distances over thirty miles.

That leaves a magnetic sense. A range of animal species, including termites, bees, and migrating birds, can detect magnetic fields. But even if pigeons have a compass sense, this cannot by itself explain homing. Imagine that you are taken to an unfamiliar place and given a compass. You will know from the compass where north is, but not where home is.

The obvious way of dealing with this problem is to postulate complex interactions between known sensory modalities, with multiple backup systems. The complex-interaction theory is safe, sounds sophisticated, and is vague enough to be irrefutable. The idea of a sense of direction involving new scientific principles is dangerous, but it may be inevitable.

A Political System Based on Empathy

SIMON BARON-COHEN

Simon Baron-Cohen is a psychologist at the Autism Research Centre, Cambridge University, and the author of *The Essential Difference: Male and Female Brains and the Truth About Autism.*

Imagine a political system based not on legal rules (systemizing) but on empathy. Would this make the world a safer place?

The British Parliament, the United States Congress, the Israeli Knesset, the French National Assembly, the Italian Senato della Repubblica, the Spanish Congreso de los Diputados – what do such political chambers have in common? Existing political systems are based on two principles: getting power through combat and then creating/revising laws and rules through combat.

Combat is sometimes physical (toppling your opponent militarily), sometimes economic (establishing a trade embargo to starve your opponent of resources), sometimes propaganda-based (waging a media campaign to discredit your opponent's reputation), and sometimes through voting-related activity (lobbying, forming alliances, fighting to win votes in key seats) with the aim of defeating the opposition.

Creating/revising laws and rules is what you do once you are in power. These might be constitutional rules, rules of precedence, judicial rulings, statutes, or other laws or codes of practice. Politicians battle for their rule-based proposal (which they hold to be best) to win and to defeat the opposition's rival proposal.

This way of doing politics is based on systemizing. First you analyze the most effective form of combat (itself a system) to win. If we do X, then we will obtain outcome Y. Then you adjust the legal code (another system). If we pass law A, we will obtain outcome B.

My colleagues and I have studied the essential difference between how men and women think. Our studies suggest that, on average, more men are systemizers and more women are empathizers. Since most political systems were set up by men, it may be no coincidence that we have ended up with political chambers built on the principles of systemizing.

So here's the dangerous new idea. What would it be like if our political chambers were based on the principles of empathizing? It is dangerous because it would mean a revolution in how we choose our politicians, how our political chambers govern, and how our politicians think and behave. We have never given such an alternative political process a chance. Might it be better and safer than what we currently have? Since empathy is about keeping in mind the thoughts and feelings of other people and not just your own, and being sensitive to another person's thoughts and feelings and not just riding roughshod over them, it is clearly incompatible with notions of 'doing battle with the opposition' and 'defeating the opposition' in order to win and hold on to power.

Currently we select party (and ultimately national) leaders based on 'leadership' qualities. Can they make decisions decisively? Can they do what is in the best interests of the party, or the country, even if it means sacrificing others to follow through on a decision? Can they ruthlessly reshuffle their cabinet and cut people loose if those people are no longer serving their interests? These are the qualities of a strong systemizer.

Note that we are not talking about whether such politicians are male or female. We are talking about how a politician (irrespective of gender) thinks and behaves.

We have had endless examples of systemizing politicians who are unable to resolve conflict. Empathizing politicians would perhaps follow Nelson Mandela's and F. W. de Klerk's example; they sat down to try to understand each other, to empathize with each other even if the other was defined as a terrorist. To do this involves the empathic act of stepping into the other's shoes and identifying with the other's feelings.

The details of a political system based on empathizing would need a lot of working out, but we can imagine certain qualities that would have no place.

Gone would be politicians who are skilled orators but simply deliver monologs, standing on a platform, pointing forcefully into the air to underline their insistence – even the body language containing an implied threat of poking their listener in the chest or the face – to win over an audience. Gone, too, would be politicians so principled they are rigid and uncompromising.

Instead, we would elect politicians based on different qualities: politicians who are good listeners, who ask questions of others instead of assuming they know the right course of action. We would instead have politicians who respond sensitively to a different point of view, and who can be flexible about where the dialog might lead. Instead of seeking to control and dominate, our politicians would be seeking to support, enable, and care.

Social Relativity

TOR NØRRETRANDERS

Tor Nørretranders is a science writer, lecturer, and consultant based in
Copenhagen, and the author of *The User Illusion: Cutting
Consciousness Down to Size* and *The Generous Man: How Helping
Others Is the Sexiest Thing You Can Do.*

Relativity is my dangerous idea. Well, neither the special nor the
general theory of relativity, but what might be called social relativity: the idea that the only thing that matters is how one
stands relative to others. That is, only the relative wealth of a
person is important. The absolute level does not matter as soon
as everyone is above a level at which his or her immediate survival needs are met.

There is now strong and consistent evidence (from fields
such as microeconomics, experimental economics, psychology,
sociolology, and primatology) that it doesn't really matter how
much you earn as long as you earn more than your wife's sister's
husband. Pioneers in these discussions have been the late British
social thinker Fred Hirsch and the American economist Robert
Frank.

But why is this idea dangerous? Because it seems to imply
that equality will never become possible in human societies:

The driving force is always to get ahead of the rest. Nobody will ever settle down and share.

So it would seem that we are forever stuck with poverty, disease, and hierarchies. This idea could make the rich and the smart lean back and forget about the rest of the pack.

But it shouldn't. Inequality may seem nice to the rich but it is not in their interest.

A huge body of epidemiological evidence suggests that inequality is in fact the prime cause of human disease. Rich people in poor countries are healthier than poor people in rich countries, even though the latter group has more resources in absolute terms. Societies with strong gradients of wealth have higher death rates and more disease, also among the people at the top. Pioneers in these studies are the British epidemiologists Michael Marmot and Richard Wilkinson.

Poverty means the spreading of disease, the degradation of ecosystems, and social violence and crime – which are also bad for the rich. Inequality means stress for everyone.

Social relativity then boils down to an illusion: It seems nice to me to be better off than the rest, but in terms of vitals – survival, good health – it is not.

Believing in social relativity can be dangerous to your health.

There *Is* Something New Under the Sun – Us

GREGORY COCHRAN

Gregory Cochran is a consultant in adaptive optics and an adjunct professor of anthropology at the University of Utah.

Thucydides said that human nature was unchanging and thus predictable, but he was probably wrong. If you consider natural selection operating in fast-changing human environments, such stasis is most unlikely. We know of a number of cases in which there has been rapid adaptive change in humans; for example, most of the malaria-defense mutations, such as sickle cell, are recent, just a few thousand years old. The lactase mutation that lets most adult Europeans digest ice cream is not much older.

There is no magic principle that restricts human evolutionary change to disease defenses and dietary adaptations: Everything is up for grabs. Genes affecting personality, reproductive strategies, cognition – all are able to change significantly over few-millennia time scales if the environment favors such change. And this includes the new environments we have made for ourselves – things like new ways of making a living and new social structures. I would be astonished if the mix of personality types favored among hunter-gatherers is exactly the same as that favored among peasant farmers ruled by a pharaoh. In fact, they might be fairly different.

There is evidence that such change has occurred. My anthropologist colleague at the University of Utah Henry Harpending and I have made a strong case that natural selection changed the Ashkenazi Jews over a thousand-year period or so, favoring certain kinds of cognitive abilities and generating genetic diseases as a side effect. The geneticist Bruce Lahn's team has found new variants of brain development genes: one, ASPM (abnormal spindle-like microcephaly associated) appears to have risen to high frequency in Europe and the Middle East in about six thousand years. We don't yet know what this new variant does, but it certainly could affect the human psyche – and if it does, Thucydides was wrong. We may not be doomed to repeat the Sicilian expedition: on the other hand, since we don't understand much yet about the changes that have occurred, we might be even more likely to be doomed. But at any rate, we have almost certainly changed. There *is* something new under the sun – us.

This concept opens strange doors. If true, it means that the people of Sumeria and Egypt's Old Kingdom were probably fundamentally different from us: Human nature has changed – some, anyhow – over recorded history. Julian Jaynes, in *The Origin of Consciousness in the Breakdown of the Bicameral Mind*, argued that there was something qualitatively different about the human mind in ancient civilization. On first reading, *Breakdown* seemed one of the craziest books ever written, but Jaynes may have been onto something.

If people a few thousand years ago thought and acted differently because of biological differences, history is never going to be the same.

A Spoon Is Like a Headache

DONALD D. HOFFMAN

Donald D. Hoffman is a cognitive scientist at the University of California at Irvine and the author of *Visual Intelligence: How We Create What We See*.

A spoon is like a headache. This is a dangerous idea in sheep's clothing. It consumes decrepit ontology, preserves methodological naturalism, and inspires exploration for a new ontology, a vehicle sufficiently robust to sustain the next leg of our search for a theory of everything.

How could a spoon and a headache do all this? Suppose I have a headache and I tell you about it. It is, say, a pounding headache that started at the back of the neck and migrated to encompass my forehead and eyes. You respond empathetically, recalling a similar headache you had, and suggest a couple of remedies. We discuss our headaches and remedies a bit, then move on to other topics.

Of course no one but me can experience my headaches, and no one but you can experience yours. But this posed no obstacle to our meaningful conversation. You simply assumed that my headaches are relevantly similar to yours, and I assumed the same about your headaches. The fact that there is no 'public headache,' no single headache that we both experience, is simply no problem.

A spoon is like a headache. Suppose I hand you a spoon. It is common to assume that the spoon I experience during this transfer is numerically identical to the spoon you experience. But this assumption is false. No one but me can experience my spoon, and no one but you can experience your spoon. But this is no problem. It is enough for me to assume that your spoon experience is relevantly similar to mine. For effective communication, no public spoon is necessary, just like no public headache is necessary. Is there a 'real spoon,' a mind-independent physical object that causes our spoon experiences and resembles our spoon experiences? This is not only unnecessary but unlikely. It is unlikely that the visual experiences of *Homo sapiens*, shaped to permit survival in a particular range of niches, should miraculously also happen to resemble the true nature of a mind-independent realm. Selective pressures for survival do not, except by accident, lead to truth.

One can have a kind of objectivity without requiring public objects. In special relativity, the measurements, and thus the experiences, of mass, length, and time differ from observer to observer, depending on their relative velocities. But these differing experiences can be related by the Lorentz transformation. This is all the objectivity one can have, and all one needs to do science.

Once we abandon public physical objects, we must reformulate many current open problems in science. One example is the mind–brain relation. There are no public brains, only my brain experiences and your brain experiences. These brain experiences are just the simplified visual experiences of *Homo sapiens*, shaped for survival in certain niches. The chances that our brain experiences resemble some mind-independent truth are remote at best, and those who would claim otherwise must surely explain the miracle. Failing a clever explanation of this miracle, there is no reason to believe that brains cause anything, including minds. And here the wolf unzips the sheepskin, and darts out into the open. The danger becomes apparent the moment we switch from boons to sprains. Oh, pardon the spoonerism.

Projection of the Longevity Curve

GERALD HOLTON

Gerald Holton is the Mallinckrodt Research Professor of Physics and Research Professor of the History of Science at Harvard University. His most recent book is *Victory and Vexation in Science: Einstein, Bohr, Heisenberg, and Others.*

Since the major absorption of scientific method into the research and practice of medicine in the 1860s, the longevity curve, at least for the white population in industrial countries, took off and has continued fairly constantly. That has been on the whole a benign result, and has begun to introduce the idea of tolerably good health as one of the basic human rights. But one now reads of projections to two hundred years and perhaps more. The economic, social, and human costs of the increasing fraction of very elderly citizens have begun to be noticed already.

To glimpse one of the possible results of the continuing projection of the longevity curve in a plausible scenario: The matriarch of the family, on her deathbed at age two hundred, is being visited by the surviving, grieving family members: a son and a daughter, each aged about 180, plus their three children, around 150 to 160 years old each, plus all their offspring, in the range of 120, 130, and so on . . . A touching picture. But what are the costs involved?

The Near-Term Inevitability of Radical Life Extension and Expansion

RAY KURZWEIL

Ray Kurzweil is an inventor and technologist. He is the author, most recently, of *The Singularity Is Near: When Humans Transcend Biology*.

My dangerous idea is the near-term inevitability of radical life extension and expansion. The idea is dangerous, however, only when contemplated from current linear perspectives.

First the inevitability: The power of information technologies is doubling each year, and moreover comprises areas beyond computation – most notably, our knowledge of biology and of our own intelligence. It took fifteen years to sequence HIV, and from that perspective the Human Genome Project seemed impossible in 1990. But the amount of genetic data we were able to sequence doubled every year, while the cost came down by half each year.

We finished the genome project on schedule, and we were able to sequence SARS in only thirty-one days. We are also gaining the means to reprogram the ancient information processes underlying biology. RNA interference can turn genes off by blocking the messenger RNA that expresses them. New forms of gene therapy are now able to place new genetic information in the right place on the right chromosome. We can create or block enzymes, the workhorses of biology. We are reverse-engineering – and gaining the means to reprogram – the information processes underlying disease and aging, and this process is accelerating, doubling every year. If we think linearly, then the idea of turning off all disease and aging processes appears far in the future, just as the genome project did in 1990. On the other hand, if we factor in the doubling of the power of these technologies each year, the prospect of radical life extension is only a couple of decades away.

In addition to reprogramming biology, we will be able to go substantially beyond biology with nanotechnology, in the form of computerized nanobots in the bloodstream. If the idea of programmable devices the size of blood cells performing thera-peutic functions in the bloodstream sounds like far-off science fiction, I would point out that we are doing this already in animals. One scientist cured type I diabetes in rats with blood-cell-sized devices containing seven nanometer pores that let insulin out in a controlled fashion and block antibodies. If we factor in the exponential advance of computation and commu-nication (price-performance multiplying by a factor of a billion in twenty-five years, while at the same time shrinking in size by a factor of thousands), these scenarios are highly realistic.

The apparent dangers are not real, while unapparent dangers *are* real. The apparent dangers are that a dramatic reduction in the death rate will create overpopulation and thereby strain energy and other resources while exacerbating environmental degradation. However, we need to capture only 1 percent of 1 percent of the sunlight to meet all our energy needs (3 percent of 1 percent by 2025) and nanoengineered solar panels and fuel cells will be able to do this, thereby meeting our energy needs in the late 2020s with clean and renewable methods. Molecular nanoassembly devices will be able to manufacture a wide range of products, just about everything we need, with inexpensive tabletop devices. The power and price-performance of these systems will double each year, much faster than the doubling rate of the biological population. As a result, poverty and pollution will decline and ultimately vanish, despite growth of the biological population.

There are real downsides, however. This is not a utopian vision. We have a new existential threat today, in the potential of a bioterrorist to engineer a new biological virus. We actually do have the knowledge to combat this problem (for example, new vaccine technologies and RNA interference, which has been shown capable of destroying arbitrary biological viruses), but it will be a race. We will have similar issues with the feasibility of self-replicating nanotechnology in the late 2020s. Containing these perils while we harvest the promise is arguably the most important issue we face.

Some people see these prospects as dangerous because they threaten their view of what it means to be human. There is a fundamental philosophical divide here. In my view, it is not our limitations that define our humanity. Rather, we are the species that seeks and succeeds in going beyond our limitations.

produced in a timely fashion. I began the school by using
what we can spare. Jamie tried a specific protocol to that
we caught a Newer animals and vegetable that not ex-
primary and human

The Domestication of Biotechnology

FREEMAN J. DYSON

Freeman J. Dyson is a theoretical physicist at the Institute for Advanced Study and the author, most recently, of *The Sun, the Genome and the Internet*.

Biotechnology will be domesticated in the next fifty years as thoroughly as computer technology was in the last fifty years.

This means cheap and user-friendly tools and do-it-yourself kits for gardeners to design their own roses and orchids and for animal breeders to design their own lizards and snakes – a new artform as creative as painting or cinema. It means biotech games for children down to kindergarten age, like computer games but played with real eggs and seeds instead of with images on a screen. Kids will grow up with an intimate feeling for the organisms they create. It means an explosion of biodiversity, as new ecologies are designed to fit into millions of local niches all over the world. Urban and rural landscapes will become more varied and more fertile.

There are two severe and obvious dangers: First, smart kids and malicious grown-ups will find ways to convert biotech tools to the manufacture of lethal microbes; second, ambitious parents will find ways to apply biotech tools to the genetic modification of their babies. The great unanswered question is whether we can regulate domesticated biotechnology so that it can be applied freely to animals and vegetables but not to microbes and humans.

Public Engagement in Science and Technology

PHILIP CAMPBELL

Philip Campbell is the editor-in-chief of *Nature*.

Scientists and governments developing public engagement about science and technology are missing the point. This turns out to be true in cases where there are collapses in consensus that have serious societal consequences. Whether in relation to climate change, genetically modified crops or Britain's triple vaccine for measles, mumps, and rubella, alternative-science networks develop among people who are neither ignorant nor irrational but have perceptions about science, the scientific literature, and its implications that differ from those prevailing in the scientific community.

Those perceptions and discussions may be half-baked but are no less powerful for all that, and they carry influence on the Internet and in the media. Researchers and governments haven't yet learned how to respond to such 'citizens' science.' Should they stop explaining and engaging? No. But they need also to understand better the influences at work within those networks (often too dismissively stereotyped) at an early stage in the debate, in order to counter bad science and minimize the impact of falsehoods.

Suppose Faulkner Was Right?

JOEL GARREAU

Joel Garreau is the cultural revolution correspondent for the *Washington Post* and the author of *Radical Evolution*.

In his December 10, 1950, Nobel Prize acceptance speech, William Faulkner said:

> I decline to accept the end of man. It is easy enough to say that man is immortal simply because he will endure: that when the last ding-dong of doom has clanged and faded from the last worthless rock hanging tideless in the last red and dying evening, that even then there will still be one more sound – that of his puny inexhaustible voice, still talking. I refuse to accept this. I believe that man will not merely endure: he will prevail.
>
> He is immortal, not because he alone among creatures has an inexhaustible voice but because he has a soul, a spirit capable of compassion and sacrifice and endurance. The poet's, the writer's, duty is to write about these things. It is his privilege to help man endure by lifting his heart, by reminding him of the courage and honor and hope and pride and compassion and pity and sacrifice which have

been the glory of his past. The poet's voice need not merely be the record of man, it can be one of the props, the pillars to help him endure and prevail.

It's easy to dismiss such optimism. The reason I hope Faulkner was right, however, is that we are at a turning point in history. For the first time, our technologies are not so much aimed outward at modifying our environment, in the fashion of fire, clothes, agriculture, cities, and space travel. Instead, they are increasingly aimed inward at modifying our minds, memories, metabolisms, personalities, and progeny. If we can do all that, then we are entering an era of engineered evolution – radical evolution, if you will – in which we take control of what it will mean to be human.

This is not some distant science fiction future. This is happening right now, in our generation, on our watch. The GRIN (genetic, robotic, information, and nano) technologies are following curves of accelerating change the arithmetic of which suggests that the last twenty years are not a guide to the next twenty. We are more likely to see that magnitude of change in the next eight. Similarly, the degree of change of the last half century, going back to the time when Faulkner spoke, may well be compressed into the next fourteen.

This raises the question of where we will gain the wisdom to guide this torrent and points to what happens if Faulkner was wrong. If we humans are unable to control our tools but instead come to be controlled by them, then we will be heading into a technodeterminist future.

You can hear different versions of what that might mean.

Some would have you believe that a future in which our creations eliminate the ills that have plagued humankind for millennia – conquering pain, suffering, stupidity, ignorance, and even death – is a vision of Heaven. Some welcome the idea that someday soon our creations will surpass the pitiful limitations of Version 1.0 humans, becoming a successor race that will conquer the universe, and care for us benevolently.

Others feel strongly that a life without suffering is a life without meaning, reducing humankind to ignominious, characterless husks. They also point to what could happen if powerful self-replicating technologies get into the hands of bumblers or madmen. They can easily imagine a vision of Hell in which we wipe out not only our species but all of life on Earth.

If Faulkner is right, however, there is a third possible future. That is the one that counts on the ragged human convoy of divergent perceptions, piqued honor, posturing, insecurity, and humor once again wending its way to glory. It puts a shocking premium on Faulkner's hope that man will prevail 'because he has a soul, a spirit capable of compassion and sacrifice and endurance.' It assumes that even as change picks up speed, giving us less and less time to react, we will still be able to rely on the impulse that Winston Churchill described when he said, 'Americans can always be counted on to do the right thing – after they have exhausted all other possibilities.'

The key measure of such a Prevail scenario's success would be an increasing intensity of links between humans, not transistors. If some sort of transcendence is achieved beyond today's understanding of human nature, it would not be through some individual becoming superman. Transcendence would be social, not solitary. The measure would be the extent to which many transform together.

The very fact that Faulkner's proposition looms so large as we look into the future does at least illuminate the present. Referring to Faulkner's breathtaking 'when the last ding-dong of doom has clanged and faded from the last worthless rock hanging tideless in the last red and dying evening, that even then there will still be one more sound: that of his puny inexhaustible voice, still talking,' the author Bruce Sterling once told me, 'You know, the most interesting part about that speech is that part . . . where William Faulkner, of all people, is alluding to H. G. Wells and the last journey of the Traveler from *The Time Machine*. It's kind of a completely heartfelt, probably drunk mishmash of cornball cryptoreligious literary humanism and the stark, bonkers, apocalyptic notions of atomic Armageddon, human extinction, and deep Darwinian geological time. Man, that was the twentieth century all over!'

What If the Unknown Becomes Known and Is Not Replaced with a New Unknown?

ERIC FISCHL

Eric Fischl is an artist in New York City (Mary Boone Gallery).

Several years ago I stood in front of a painting by Vermeer. It was a painting of a woman reading a letter. She stood near the window for better lighting and behind her hung a map of the known world. I was stunned by the revelation of this work. Vermeer understood something so basic to human need that it had gone virtually unnoticed: communication from afar.

Everything we have done to make us more capable, more powerful, better protected, more intelligent has been done by overcoming our physical limitations and enhancing our perceptual abilities and our adaptability. When I think of Vermeer's woman reading the letter, I wonder how long it took to get to her? Then I think, My god, at some time we developed a system in which one could leave home and send word back! We figured out a way to be heard from far away and then developed another system so that we could be seen from far away. Then I start to marvel at the alchemy of painting and how we have been able to invest materials with consciousness so that Vermeer can talk to me across time. I see, too, that he has put me in the position of not knowing, as I am kept from reading the contents of the letter. In this way, he has placed me at the edge, the frontier of wanting to know what I cannot know. I want to know how long this letter-sender has been away and what he has been doing all this time. Is he safe? Does he still love her? Is he on his way home?

Vermeer puts me into what had been her condition of uncertainty. All I can do is wonder and wait. This makes me think about how not knowing is so important. Not knowing makes the world large and uncertain and our survival tenuous. It is a mystery why humans roam, and still more a mystery why we need to feel so connected to the place we've left. The not knowing causes such profound anxiety that it in turn spawns creativity. The impetus for this creativity is empowerment. Our gadgets, gizmos, networks of transportation and communication have all been developed either to explore, utilize, or master the unknown territory.

If the unknown becomes known and is not replaced with a new unknown, if the farther we reach outward is connected only to how fast we can bring it home, if the time between not knowing and knowing becomes too small, creativity will falter. And so I worry that if we bring the universe more completely, more effortlessly, into our homes, there will be less and less reason to leave them.

Where Goods Cross Frontiers, Armies Won't

MICHAEL SHERMER

Michael Shermer is the publisher of *Skeptic* magazine, a monthly columnist for *Scientific American*, and the author most recently of *Science Friction: Where the Known Meets the Unknown*.

Where goods cross frontiers, armies won't. Restated: Where economic borders between two nations are porous, political borders become impervious to armies.

Data from the new sciences of evolutionary economics, behavioral economics, and neuroeconomics reveal that when people are free to cooperate and trade (such as in game-theory protocols), they establish trust, reinforced through neural pathways that release such bonding hormones as oxytocin. Thus, biologically they are less likely to fight and kill those with whom they are cooperating and trading.

My dangerous idea is a solution to what I call *the really hard problem*: How best should we live? My answer: In a free society – defined as free-market economics and democratic politics (fiscal conservatism and social liberalism), leading to the greatest liberty for the greatest number. Since humans are by nature tribal, the overall goal is to expand the concept of the tribe to include *all* members of the species, in a global free society. Free trade between all peoples is the surest way to reach this goal.

People have a hard time accepting free-market economics for the same reason that they have a hard time accepting evolution: It is counterintuitive. Life looks intelligently designed, so our natural inclination is to infer that there must be an intelligent designer – a God. Similarly, the economy looks designed, so our natural inclination is to infer that we need a designer – a Government. In fact, complexity theory explains how the principles of self-organization and emergence cause complex systems to arise from simple systems without a top-down designer.

Charles Darwin's natural selection is Adam Smith's invisible hand. Darwin showed how complex design and ecological balance were unintended consequences of individual competition among organisms. Smith showed how national wealth and social harmony were unintended consequences of individual competition among people. Nature's economy mirrors society's economy. Thus, integrating evolution and economics – what I call *evonomics* – demonstrates that an old economic doctrine is supported by modern biology.

Government Is the Problem, Not the Solution

MATT RIDLEY

Matt Ridley is a science writer and the founding chairman of the International Centre for Life. He is the author, most recently, of *Francis Crick: Discoverer of the Genetic Code.*

In all times and in all places there has been too much government. We now know what prosperity is: It is the gradual extension of the division of labor through the free exchange of goods and ideas, and the consequent introduction of efficiencies by the invention of new technologies. This is the process that has given us health, wealth, and wisdom on a scale unimagined by our ancestors. It not only raises material standards of living, it also fuels social integration, fairness, and charity. It has never failed yet. No society has grown poorer or more unequal through trade, exchange, and invention. Think of pre-Ming as opposed to Ming China, seventeenth-century Holland as opposed to imperial Spain, eighteenth-century England as opposed to Louis XIV's France, twentieth-century America as opposed to Stalin's Russia, and postwar Japan, Hong Kong, and Korea as opposed to Ghana, Cuba, and Argentina. Think of the Phoenicians as opposed to the Egyptians, Athens as opposed to Sparta, the Hanseatic League as opposed to the

Roman Empire. In every case, weak or decentralized government but strong free trade led to surges in prosperity for all, whereas strong central government led to parasitic, tax-fed officialdom, a stifling of innovation, relative economic decline, and usually war.

Take Rome. It prospered because it was a free trade zone. But it repeatedly invested the proceeds of that prosperity in too much government and so wasted it in luxury, war, gladiators, and public monuments. The Roman Empire's list of innovations is derisory, even compared with that of the Dark Ages that followed.

In every age and at every time, there have been people who say we need more regulation, more government. Sometimes they say we need it to protect exchange from corruption, to set the standards and police the rules – in which case they have a point, though often they exaggerate it. Self-policing standards and rules were developed by free-trading merchants in medieval Europe long before those standards and rules were taken over and codified as laws (and often corrupted) by monarchs and governments.

Sometimes they say we need it to protect the weak, or the victims of technological change or trade flows. But throughout history such intervention, though well meant, has usually failed those it intended to help, because its progenitors refuse to believe in (or find out about) David Ricardo's Law of Comparative Advantage: Even if China is better at making everything than France is, there will still be a million things it pays China to buy from France rather than make itself. Why? Because rather than invent, say, luxury goods or insurance services, China will find that it pays to make more T-shirts and use the proceeds to import luxury goods and insurance.

Government is a very dangerous toy. It is used to fight wars, impose ideologies, and enrich rulers. True, nowadays our leaders do not usually enrich themselves (at least not on the scale of the Sun King), but they enrich their clients: They preside over vast and insatiable parasitic bureaucracies that grow by Parkinson's law and live off true creators of wealth, such as traders and inventors.

Sure, it is possible to have too little government. Except that has not been the world's problem for millennia. After the century of Mao, Hitler, and Stalin, can anybody really say that the risk of too little government is greater than the risk of too much? Or that it is Africa's problem today? The dangerous idea we all need to learn is that the more we limit the growth of government, the better off we will all be.

The Free Market

MIHALY CSIKSZENTMIHALYI

Mihaly Csikszentmihalyi is a psychologist and the director of the Quality of Life Research Center at Claremont Graduate University. He is the author, most recently, of *Good Business: Leadership, Flow, and the Making of Meaning*.

Generally ideas are thought to be dangerous when they threaten an entrenched authority. Galileo was sued not because he claimed that the earth revolved around the sun (a 'hypothesis' his chief prosecutor, Cardinal Bellarmine, apparently was quite willing to entertain in private) but because the Church could not tolerate reversal by another epistemology – in this case, the scientific method. Similar conflicts arose when Darwin's view of how humans first appeared on the planet challenged religious accounts of creation, and when Mendelian genetics applied to the growth of hardier strains of wheat challenged Leninist doctrine as interpreted by Lysenko.

One of the most dangerous ideas at large in the current culture is that the 'free market' is the ultimate arbiter of political decisions, and that there is an 'invisible hand' that will direct us to the most desirable future, provided the free market is allowed to actualize itself. This mystical faith is based on reasonable empirical foundations, but when it is embraced as a final solution to the ills of humankind, it risks destroying both the material resources and the cultural achievements our species has so painstakingly developed.

So the dangerous idea on which our culture is based is that the political economy has a silver bullet – the free market – that must take precedence over any other value and will thereby lead to peace and prosperity. It is dangerous because, like all silver bullets, it is an intellectual and political scam that might benefit some but ultimately requires the majority to pay for the destruction it causes.

My dangerous idea is dangerous only to those who support the hegemony of the market. It consists in pointing out that the imperial free market wears no clothes: It does not exist in the first place, and what passes for it is dangerous to the well-being of our species. Scientists need to turn their attention to what the complex system that is human life will require in the future.

Beginnings like the Calvert-Henderson 'Quality of Life Indicators,' which focus on such central requirements as health, education, infrastructure, environment, human rights, and public safety, need to become part of our social and political agenda. And when their findings come into conflict with the agenda of the prophets of the free market, the conflict should be examined. Who is it that benefits from the erosion of the quality of life?

Modern Science Is a Product of Biology

ARNOLD TREHUB

Arnold Trehub is an adjunct professor of psychology at the University of Massachusetts, Amherst, and the author of *The Cognitive Brain*.

The entire conceptual edifice of modern science is a product of biology. Even the most basic and profound ideas of science – relativity, quantum theory, the theory of evolution by natural selection – are generated and necessarily limited by the particular capacities of our human biology. This implies that the content and scope of scientific knowledge is not open-ended.

No More Teacher's Dirty Looks

ROGER C. SCHANK

Roger C. Schank is a cognitive psychologist, computer scientist, professor emeritus at Northwestern, chief learning officer at Trump University, and the author, most recently, of *Lessons in Learning, e-learning, and Training: Perspectives and Guidance for the Enlightened Trainer.*

After a natural disaster, the newscasters eventually excitedly announce that school is finally open, so no matter what else is terrible where they live, the kids are going to school. I always feel sorry for the poor kids.

My dangerous idea is one that most people immediately reject without giving it serious thought: School is bad for kids. It makes them unhappy and, as tests show, they don't learn much.

When you listen to children talk about school, you easily discover what they are thinking about in school: Who likes them, who is being mean to them, how to improve their social ranking, how to get the teacher to treat them well and give them good grades.

Schools are structured today in much the same way as they have been for hundreds of years. And for hundreds of years, philosophers and others have pointed out that school is really a bad idea:

We are shut up in schools and college recitation rooms for ten or fifteen years, and come out at last with a belly full of words and do not know a thing. – Ralph Waldo Emerson

Education is an admirable thing, but it is well to remember from time to time that nothing that is worth knowing can be taught. – Oscar Wilde

Schools should simply cease to exist as we know them. The government needs to get out of the education business and stop thinking that it knows what children should know and then testing them constantly to see if they regurgitate whatever they have just been spoonfed.

The government is and always has been the problem in education:

If the government would make up its mind to require for every child a good education, it might save itself the trouble of providing one. It might leave to parents to obtain the education where and how they pleased, and content itself with helping to pay the school fees of the poorer classes of children, and defraying the entire school expenses of those who have no one else to pay for them. – J. S. Mill

First, God created idiots. That was just for practice. Then He created school boards. – Mark Twain

Schools need to be replaced by safe places where children can go to learn how to do things they are interested in learning how to do. Their interests should guide their learning. The government's role should be to create places that are attractive to children and would cause them to want to go there.

Whence it comes to pass, that for not having chosen the right course, we often take very great pains, and consume a good part of our time in training up children to things, for which, by their natural constitution, they are totally unfit. – Montaigne

There are two types of education . . . One should teach us how to make a living, and the other how to live. – John Adams

Over a million students have opted out of the existing school system and are now being home-schooled. The problem is that the states regulate home-schooling, and home-schooling still looks an awful lot like school.

We need to stop producing a nation of stressed-out students who learn how to please the teacher instead of pleasing themselves. We need to produce adults who love learning, not adults who avoid all learning because it reminds them of the horrors of school. We need to stop thinking that all children need to learn the same stuff. We need to create adults who can think for themselves and are not convinced about how to understand complex situations in simplistic terms that can be rendered in a sound bite.

Just call school off. Turn them all into recreation centers.

We Are All Virtual

CLIFFORD PICKOVER

Clifford Pickover is a computer scientist and a staff member at the IBM
T. J. Watson Research Center. He is the author of numerous books,
including, most recently, *Sex, Drugs, Einstein, and Elves: Sushi,
Psychedelics, Parallel Universes, and the Quest for Transcendence.*

Our desire to experience entertaining virtual realities is
increasing. As our understanding of the human brain also
accelerates, we will create both imagined realities and a set of
memories to support these simulacrums. For example, someday it will be possible to simulate your visit to the Middle
Ages and, to make the experience realistic, we may wish to
ensure that you *believe* yourself to actually be in the Middle
Ages. False memories may be implanted, temporarily overriding your real memories. This should be easy to do, given
that we can already coax the mind to create richly detailed
virtual worlds filled with ornate palaces and strange beings
through the use of the drug DMT (dimethyltryptamine).
The brains of people who take DMT seem able to access a
treasure chest of images and experiences, which typically
include jeweled cities and temples, angelic beings, feline
shapes, serpents, and shiny metals. When we understand the

brain better, we will be able to safely generate more controlled visions.

Our brains are also capable of simulating complex worlds when we dream. For example, after I watched a movie about people in a coastal town during the Renaissance, I was transported there later that night, in a dream. The mental simulation of the Renaissance did not have to be perfect, and I'm sure that there were myriad flaws; however, during that dream I *believed* I was in the Renaissance. If we understood the nature of how the mind induces the conviction of reality even when strange, nonphysical events happen in dreams, we could use this knowledge to ensure that your simulated trip to the Middle Ages seemed utterly real even if the simulation was imperfect. It will be easy to create seemingly realistic virtual realities, because the accuracy of our simulations need not be perfect, or even good, to make them seem real. After all, our nightly dreams usually seem quite real, despite the logical or structural inconsistencies we recognize on awakening.

In the future, you will personally create ten simulated lives. Your day job is a computer programmer for IBM. However, after work, you'll be a knight in shining armor in the Middle Ages, attending lavish banquets, smiling at wandering minstrels and beautiful princesses. The next night, you'll be in the Renaissance, living in your home on the Amalfi coast of Italy, enjoying a dinner of plover, pigeon, and heron. If this ratio of one real life to ten simulated lives turns out to be representative of human experience, it means that, right now, you have only one in ten chances of actually being alive in the actual present.

Runaway Consumerism Explains the Fermi Paradox

GEOFFREY MILLER

Geoffrey Miller is an evolutionary psychologist at the University of New Mexico and the author of *The Mating Mind: How Sexual Choice Shaped the Evolution of Human Nature*.

The story goes like this: Sometime in the 1940s, Enrico Fermi was talking about the possibility of extraterrestrial intelligence with some other physicists. They were impressed that our galaxy holds 100 billion stars, that life evolved quickly and progressively on Earth, and that an intelligent, exponentially reproducing species could colonize the galaxy in just a few million years. They reasoned that extraterrestrial intelligence should be common by now. Fermi listened patiently, then asked simply, 'So, where is everybody?' That is, if extraterrestrial intelligence is common, why haven't we met any bright aliens yet? This conundrum became known as Fermi's paradox.

The paradox has become ever more baffling. More than a hundred and fifty extrasolar planets have been identified in the last few years, suggesting that life-hospitable planets orbit most stars. Paleontology shows that organic life evolved very quickly after Earth's surface cooled and became hospitable to it. Given simple life, evolution shows a progressive trend toward larger bodies, brains, and social complexity. Evolutionary psychology reveals several credible paths from simpler social minds to human-level creative intelligence. Yet forty-some years of intensive searching for extraterrestrial intelligence have yielded nothing. No radio signals, no credible spacecraft sightings, no close encounters of any kind.

So it looks as if there are two possibilities. Perhaps our science overestimates the likelihood of extraterrestrial intelligence evolving. Or perhaps evolved technical intelligence has some deep tendency to be self-limiting, even self-exterminating. After Hiroshima, some suggested that any aliens bright enough to make colonizing spaceships would be bright enough to make thermonuclear bombs and would use them on one another sooner or later. Perhaps extraterrestrial intelligence always blows itself up. Fermi's paradox became, for a while, a cautionary tale about cold-war geopolitics.

I suggest a different, even darker solution to Fermi's paradox. Basically I think aliens don't blow themselves up; they just get addicted to computer games. They forget to send radio signals or colonize space because they're too busy with runaway consumerism and virtual-reality narcissism. They don't need Sentinels to enslave them in a Matrix; they do it to themselves, just as we are doing today.

The fundamental problem is that any evolved mind must pay attention to indirect cues of biological fitness rather than tracking fitness itself. We don't seek reproductive success directly; we have always sought tasty foods, which tended to promote survival, and luscious mates, who tended to produce bright, healthy babies. Modern results: fast food and pornography. Technology is fairly good at controlling external reality to promote our real biological fitness, but it's even better at delivering fake fitness – subjective cues of survival and reproduction without the real-world effects. Fresh organic fruit juice costs so much more than nutrition-free soda. Having real friends is so much more of an effort than watching *Friends* on TV. Actually colonizing the galaxy would be so much harder than pretending to have done so by filming *Star Wars*.

Fitness-faking technology tends to evolve much faster than our psychological resistance to it. The printing press is invented; people read more novels and have fewer kids; only a few curmudgeons lament this. The Xbox 360 is invented; people would rather play a high-resolution virtual ape in Peter Jackson's *King Kong* than be a perfect-resolution real human. Teens today must find their way through a carnival of addictively fitness-faking entertainment products: MP3, DVD, TiVo, XM radio, Verizon cellphones, Spice cable, EverQuest online, instant messaging, Ecstasy, BC Bud. The traditional staples of physical, mental, and social development (athletics, homework, dating) are neglected. The few young people with the self-control to pursue the meritocratic path often get distracted at the last minute: The MIT graduates apply to do computer-game design for Electronics Arts rather than rocket science for NASA.

Around 1900, most inventions concerned physical reality: cars, airplanes, zeppelins, electric lights, vacuum cleaners, air conditioners, bras, zippers. In 2005, most inventions concern virtual entertainment – the top ten patent recipients are usually IBM, Matsushita, Canon, Hewlett-Packard, Micron Technology, Samsung, Intel, Hitachi, Toshiba, and Sony – not Boeing, Toyota, or Wonderbra. We have already shifted from a reality economy to a virtual economy, from physics to psychology as the value driver and resource allocator. We are already disappearing up our own brainstems. Freud's pleasure principle triumphs over the reality principle. We narrowcast human interest stories to one another rather than broadcasting messages of universal peace and progress to other star systems.

Maybe the bright aliens did the same. I suspect that a certain period of fitness-faking narcissism is inevitable after any intelligent life evolves. This is the Great Temptation for any technological species – to shape their subjective reality to provide the cues of survival and reproductive success without the substance. Most bright alien species probably go extinct gradually, allocating more time and resources to their pleasures and less to their children.

Heritable variation in personality might allow some lineages to resist the Great Temptation and last longer. Those who persist will evolve more self-control, conscientiousness, and pragmatism. They will evolve a horror of virtual entertainment, psychoactive drugs, and contraception. They will stress the values of hard work, delayed gratification, child-rearing, and environmental stewardship. They will combine the family values of the religious right with the sustainability values of the Greenpeace left.

My dangerous idea-within-an-idea is that this, too, is already happening. Christian and Muslim fundamentalists and anti-consumerism activists already understand exactly what the Great Temptation is and how to avoid it. They insulate themselves from our Creative-Class dreamworlds and our EverQuest economics. They wait patiently for our fitness-faking narcissism to go extinct. Those practical-minded breeders will inherit the earth, as likeminded aliens may have inherited a few other planets. When they finally achieve Contact, it will not be a meeting of novel readers and game players. It will be a meeting of dead-serious superparents, who congratulate one another on surviving not just the Bomb but the Xbox. They will toast one another not in a soft-porn Holodeck but in a sacred nursery.

Simulation Versus Authenticity

SHERRY TURKLE

Sherry Turkle is Abby Rockefeller Mauzé Professor of the Social Studies of Science and Technology in the Program in Science, Technology, and Society at MIT and the founder and director of the MIT Initiative on Technology and Self. She is the author of *The Second Self: Computers and the Human Spirit.*

Consider this moment from 2005: I take my fourteen-year-old daughter to the Darwin exhibit at the American Museum of Natural History. The exhibit documents Darwin's life and thought, and with a somewhat defensive tone (in light of current challenges to evolution by proponents of intelligent design) presents the theory of evolution as the central truth underpinning contemporary biology. The Darwin exhibit wants to convince and it wants to please. At the entrance to the exhibit is a turtle from the Galápagos Islands, a seminal object in the development of evolutionary theory. The turtle rests in its cage, utterly still. 'They could have used a robot,' comments my daughter. It was a shame to bring the turtle all this way and put it in a cage for a performance that draws so little on the turtle's 'aliveness.'

I am startled by her comments, both solicitous of the

imprisoned turtle because it is alive and unconcerned by its authenticity. The museum has been advertising these turtles as wonders, curiosities, marvels – among the plastic models of life at the museum, here is the life that Darwin saw. I begin to talk with others at the exhibit, parents and children. It is Thanksgiving weekend. The line is long, the crowd frozen in place. My question, 'Do you care that the turtle is alive?' is welcome diversion. A ten-year-old girl would prefer a robot turtle because aliveness comes with aesthetic inconvenience: 'Its water looks dirty. Gross!' More usually, the votes for the robots echo my daughter's sentiment that in this setting aliveness doesn't seem worth the trouble. A twelve-year-old girl opines: 'For what the turtles do, you didn't have to have the live ones.' Her father looks at her, uncomprehending: 'But the point is that they are real, that's the whole point.'

The Darwin exhibit is about authenticity: On display are the actual magnifying glass that Darwin used, the actual notebooks in which he recorded his observations – indeed, the very notebook in which he wrote the famous sentences that first described his theory of evolution. But in the children's reactions to the inert but alive Galápagos turtle, the idea of the 'original' is in crisis.

I have long believed that in the culture of simulation, the notion of authenticity is for us what sex was to the Victorians – 'threat and obsession, taboo and fascination.' I have lived with this idea for many years, yet at the museum I find the children's position startling, strangely unsettling. The dangerous idea is that for a new generation, simulation is not connoted as second best. For these children, in this context, aliveness seems to have no intrinsic value. Rather, it is useful only if needed for a specific purpose. 'If you put in a robot instead of the live turtle, do you think people should be told that the turtle is not alive?' I ask. Not really, say several of the children. Data on 'aliveness' can be shared on a 'need to know' basis, for a purpose. But what *are* the purposes of living things? When do we need to know whether or not something is alive?

Consider another vignette from 2005: An elderly woman in a nursing home outside of Boston is sad. Her son has broken off his relationship with her. Her nursing home is part of a study I am conducting on robotics for the elderly. I am recording her reactions as she sits with the robot Paro, a seal-like creature, advertised as the first 'therapeutic robot' for its ostensibly positive effects on the ill, the elderly, and the emotionally troubled. Paro is able to make eye contact through sensing the direction of a human voice, is sensitive to touch, and has 'states of mind' that are affected by how it is treated, for example, is it stroked gently or aggressively? In this session with Paro, the woman, depressed because of her son's abandonment, comes to believe that the robot is depressed as well. She turns to Paro, strokes him, and says: 'Yes, you're sad, aren't you? It's tough out there. Yes, it's hard.' And then she pets the robot once again, attempting to provide it with comfort. And in so doing, she tries to comfort herself.

The woman's sense of being understood is based on the ability of computational objects like Paro to convince their users that they are in a relationship. I call these creatures (some virtual, some physical robots) relational artifacts. Their ability to inspire relationship is not based on their intelligence or consciousness but on their ability to push certain Darwinian buttons in people (making eye contact, for example) that make people respond *as though* they were in relationship. For me, relational artifacts are the new uncanny in our computer culture – as Freud once put it, the long familiar taking a form that is strangely unfamiliar. As such, they confront us with new questions.

What does this deployment of 'nurturing technology' at the two most dependent moments of the life cycle say about us? What will it do to us? Do plans to provide relational robots to attend to children and the elderly make us less likely to look for other solutions for their care? People come to feel love for their robots, but if our experience with relational artifacts is based on a fundamentally deceitful interchange, can it be good for us? Or might it be good for us in the feel-good sense, but bad for us in our lives as moral beings?

Relationships with robots bring us back to Darwin and *his* dangerous idea: the challenge to human uniqueness. When we see children and the elderly exchanging tendernesses with robotic pets, the most important question is not whether children will love their robotic pets more than their real-life pets or even their parents but rather what will 'loving' come to mean?

Culture Is Natural

DAN SPERBER

Dan Sperber is a social and cognitive scientist at the Centre National de la Recherche Scientifique, Paris, and the author of *Explaining Culture: A Naturalistic Approach.*

A number of us – biologists, cognitive scientists, anthropologists, philosophers – have been trying to lay the foundations for a truly naturalistic approach to culture. Sociobiologists and cultural ecologists have explored the idea that cultural behaviors are biological adaptations, to be explained in terms of natural selection. Memeticists, inspired by the evolutionary biologist Richard Dawkins, argue that cultural evolution is an autonomous, Darwinian selection process enabled, but not governed, by biological evolution.

The evolutionary psychologists Luca Cavalli-Sforza, Mark Feldman, Robert Boyd, Peter Richerson, and I are among those who, in different ways, argue for interactions between biology and culture that are more complex. These naturalistic approaches have been received not just with intellectual objections but also with moral and political outrage: This is a dangerous idea, to be strenuously resisted, for it threatens humanistic values and sound social sciences.

When I am called a reductionist, I take it as an unearned compliment. A genuine reduction is a great scientific achievement, but – too bad – the naturalistic study of culture I advocate does not to reduce to biology or psychology. When I am called a positivist (an insult among postmodernists), I acknowledge without any sense of guilt or inadequacy that indeed I don't believe that all facts are socially constructed. On the whole, having one's ideas described as dangerous is flattering.

Dangerous ideas are potentially important. Braving insults and misrepresentations in defending these ideas is noble. Many advocates of naturalistic approaches to culture see themselves as a group of free-thinking, deep-probing scholars besieged by bigots.

But wait a minute! Naturalistic approaches *can* be dangerous: after all, they have been. The use of biological evidence and arguments purporting to show profound natural inequalities among human 'races' or ethnic groups, or between women and men, is only too well represented in the history of our disciplines. It is not good enough for us to point out (rightly) that (1) the science involved is bad science; (2) even if some natural inequality were established, it would not come near justifying any inequality in rights; and (3) postmodernists criticizing naturalism on political grounds should begin by rejecting Heidegger and other reactionaries in their pantheon who also have been accomplices of policies of discrimination. This is not enough because the racist and sexist uses of naturalism are not exactly unfortunate accidents.

Species evolve because of genetic differences among their members; therefore you cannot leave biological difference out of a biological approach. Luckily, it so happens that biological differences among humans are minor and don't produce subspecies or 'races,' and human sexual dimorphism is relatively limited. In particular, all humans have minds/brains made up of the same mechanisms, with just fine-tuning differences. (Think how very different all this would be if, however improbably, Neanderthals had survived and developed culturally as we did, so that there really were different human races.)

Given what anthropologists have long called 'the psychic unity of the human kind,' the fundamental goal for a naturalistic approach is to explain how a common human nature – and not biological differences among humans – gives rise to such a diversity of languages, cultures, and social organizations. Given the real and present danger of distortion and exploitation, it must be part of our agenda to take responsibility for the way this approach is understood by a wider public.

This, happily, has been done by a number of outstanding authors capable of explaining serious science to lay audiences and who typically have warned their readers against misuses of biology. So the danger is being averted, and let's just move on? No, we are not there yet, because the very necessity of popularizing the naturalistic approach and the very talent with which this is being done creates a new danger – that of arrogance.

We naturalists do have radical objections to what Leda Cosmides and John Tooby have called the 'standard social science model.' We have many insightful hypotheses and even some relevant data. The truth of the matter, however, is that naturalistic approaches to culture have so far remained speculative, only beginning to throw light on fragments of the extraordinarily wide range of detailed evidence accumulated by historians, anthropologists, sociologists, and others. Many of those who find our ideas dangerous fear what they see as an imperialistic bid to take over their domain.

The bid would be unrealistic, and so is the fear. The real risk is different. The social sciences host a variety of approaches, which, with a few high-profile exceptions, contribute to our understanding of the domain. Even if it involves some reshuffling, a naturalistic approach should be seen as a particularly welcome and important addition. But naturalists full of grand claims and promises but with little interest in the competence accumulated by others are, if not exactly dangerous, at least much less useful than they should be, and the deeper challenge they present to social scientists' mental habits is less likely to be properly met.

The Human Brain Is a Cultural Artifact

TIMOTHY TAYLOR

Timothy Taylor is an archaeologist at the University of Bradford in the United Kingdom and the author of *The Buried Soul: How Humans Invented Death*.

Phylogenetically, humans are an evolutionary puzzle. Walking on two legs freed the hands to do new things, like chip stones to make tools – the first artifacts, dating to 2.7 million years ago – but it also narrows the pelvis, limiting the possible size of the fetal cranium. Thus the brain expansion that began 2 million years ago should not have happened.

But imagine that, along with making chipped-stone tools, one genus of hominid appropriates the looped entrails of a dead animal, or learns to tie a simple knot, and invents a sling. (Chimpanzees are known to carry water in leaves and gorillas to measure water depth with sticks, so the practical and abstract thinking required here can be safely assumed for our human ancestors by this point.)

In its sling, the hominid child can now hip-ride with little impairment to its parent's hands-free movement. This has the unexpected and certainly unplanned consequence that it is no longer important for the infant to be able to hang on, as chimps do. Although because of the biomechanical constraints of a bipedal pelvis the hominid child cannot be *born* with a big head (and thus with a large initial brain capacity), it can now be born underdeveloped. That is to say, the sling allows fetuses to be born in an ever more ontogenically retarded state. This trend, which humans do indeed display, is called neoteny. The retention of earlier features for a longer time means that the total developmental sequence extends far beyond the nine months of natural gestation. Hominid children, born under-developed, could grow their crania *outside* the womb, in the pseudomarsupial pouch of an infant-carrying sling.

From this point onward, it is not hard to see how a distinctively human culture emerges through the extrauterine formation of higher cognitive capacities – the phylogenetic and ontogenic icing on the cake of primate brain function. The child, carried by the parent into social situations, watches vocalization. Parental selection for smart features, such as an ability to babble early, may well, as others have suggested, have driven the brain size increases up until two hundred and fifty thousand years ago – the point when the final biomechanical limits of big-headed mammals with narrow pelvises were reached by two species, Neanderthals and us.

This is the phylogenetic side of the case. In terms of ontogeny, the obvious applies: It recapitulates phylogeny. The underdeveloped brains of hominid infants were culture-prone, and in this sense I do not dissent from Dan Sperber's dangerous idea that 'culture is natural.' But human culture – unlike the basic culture of learned routines and tool-using observed in various mammals – is a system of signs, essentially the association of words with things and the ascription and recognition of value in relation to this.

As the philosopher and social anthropologist Ernest Gellner has pointed out, human beings, taken cross-culturally as a species, exhibit by far the greatest range of behavioral variation of any animal. However, within any ongoing community of people possessing language, ideology, and a culturally inherited and developed technology, conformity has usually been a paramount value, with death often the price for dissent. My belief is that because of the malleability of the neotenic brain, *cultural systems are physically built into the developing tissue of the mind.*

Instead of seeing the brain as the genetic hardware into which the cultural software is loaded and then arguing about the relative determining influences of each in areas such as, say, sexual orientation or mathematical ability (the old nature-nurture debate), we can conclude that culture (as the evolutionary biologist Richard Dawkins long ago noted, with respect to contraception) acts to subvert genes but is also enabled by them. Ontogenic retardation allowed both environment and the developing milieu of cultural routines to act on brain-hardware construction, alongside the working through of the genetic blueprint. The fact that the modern human brain is coded for by genes does not mean that the critical self-consciousness for which it is famous (in its own community of brains) is noncultural, any more than a barbed and tanged arrowhead is noncultural just because it is made of flint.

The human brain is able to go not just beyond nature but beyond culture, too, by dissenting from old norms and establishing others. The emergence of the high arts and science is part of this process of the human brain, with its instrumental extrasomatic adaptations and memory stores (books, laboratories, computers), and is underpinned by the most critical development in the encultured human brain: free will.

However, not all humans – or all human communities – seem capable of equal levels of free will. In extreme cases, they appear to display none at all. Reasons include genetic incapacity, but it is also possible for a lack of mental freedom to be culturally engendered and sometimes even encouraged. Archaeologically, the evidence is there from the first farming societies in Europe: the Neolithic massacre at Talheim, where an entire community was genocidally wiped out except for the youngest children, has been taken as evidence (supported by anthropological analogies) of the re-enculturation of still flexible minds within the community of the victors to serve and live out their orphaned lives as slaves. In the future, one might surmise that the dark side of the development of virtual reality machines (described in these pages by Clifford Pickover) will be the infinitely more subtle cultural programming of impressionable individuals as sophisticated conformists.

The interplay of genes and culture has produced in us potential for a formidable range of abilities and intelligences. It is critical that in the future we both fulfill and extend this potential in the realm of judgment, choice, and understanding in both sciences and arts. But the idea of the brain as a cultural artifact is dangerous. Those with an interest in social engineering – tyrants and authoritarian regimes – will almost certainly attempt to develop it to their advantage. Free will is threatening to the powerful who, by understanding its formation, will act to undermine it in sophisticated ways. The usefulness of cultural artifacts that have the degree of complexity of human brains makes our own species the most obvious candidate for the enhanced super-robot of the future – not just smart factory operatives and docile consumers but cunning weapons delivery systems (suicide bombers) and conformity enforcers. At worst, the very special qualities of human life that have been enabled by our remarkable natural history, the confluence of genes and culture, could end up as a realm of freedom for an elite few.

Free Will Is Exercised Unconsciously

ERIC R. KANDEL

Eric R. Kandel is a biochemist and University Professor at Columbia University. The essay below is taken from his recent book, *In Search of Memory: The Emergence of a New Science of Mind.*

Sigmund Freud emphasized at the beginning of the twentieth century that most of our perceptual and cognitive processes are unconscious – except those that are in the immediate focus of our attention – and that unconscious mental processes guide much of human behavior.

Freud's idea was a natural extension of the notion of *unconscious inference,* proposed in the 1860s by Hermann Helmholtz, the German physicist turned neural scientist. Helmholtz was the first to measure the conduction of electrical signals in nerves. He had expected it to be as the speed of light, as fast as the conduction of electricity in copper cables, and found to his surprise that it was much slower, only about 90m sec. He then examined the reaction time – the time it takes a subject to respond to a consciously perceived stimulus, and found that it was much, much slower than even the combined conduction times required for sensory and motor activities.

This caused Helmholz to argue that a great deal of brain

processing occurred unconsciously prior to the conscious perception of an object, that much of what goes on in the brain is
not represented in consciousness, and that the perception of
objects depends on 'unconscious inferences' made by the brain,
based on thinking and reasoning without awareness. This view
was not accepted by many brain scientists, who believed that
consciousness is necessary for making inferences. However, in
the 1970s a number of experiments began to accumulate in
favor of the idea that most cognitive processes that occur in the
brain never enter consciousness.

Perhaps the most influential of these were carried out by
Benjamin Libet in 1986. Libet used as his starting point a discovery made by the German neurologist Hans Kornhuber.
Kornhuber asked volunteers to move their right index finger.
He then measured this voluntary movement with a strain gauge
while at the same time recording the electrical activity of the
brain by means of an electrode on the skull. After hundreds of
trials, Kornhuber found that invariably each movement was
preceded by a little blip in the electrical record from the brain,
a spark of free will! He called this potential in the brain the
'readiness potential' and found that it occurred one second
before the voluntary movement.

Libet followed up on Kornhuber's finding with an experiment in which he asked volunteers to lift a finger whenever they
felt the urge to do so. He placed an electrode on a volunteer's
skull and confirmed a readiness potential about one second
before the person lifted his or her finger. He then compared the
time it took for the person to will the movement with the time
of the readiness potential. Amazingly, he found that the readiness potential appeared not after but 200 milliseconds before a
person felt the urge to move his or her finger! Thus by merely

observing the electrical activity of the brain, Libet could predict what the subject would do before the subject was aware of having decided to do it.

These experiments have caused philosophers of mind to ask: If the choice is determined in the brain unconsciously before we decide to act, where is free will?

Are these choices predetermined? Is our experience of freely willing our actions only an illusion, a rationalization after the fact? Freud, Helmholtz, and Libet would disagree and argue that the choice is freely made but that it is made without our awareness. Libet, for example, proposes that the process of initiating a voluntary action occurs in an unconscious part of the brain but that just before the action is initiated, consciousness is recruited to approve or veto the action. In the 200 milliseconds before a finger is lifted, consciousness determines whether it moves or not.

Whatever the reasons for the delay between decision and awareness, Libet's findings now raise the moral question: Is one to be held responsible for decisions that are made without conscious awareness?

Free Will Is Going Away

CLAY SHIRKY

Clay Shirky is an adjunct professor in New York University Graduate School's Interactive Telecommunications Program and the author of *Voices from the Net*.

In 2002, a group of teenagers sued McDonald's for making them fat, charging, among other things, that McDonald's used promotional techniques to get them to eat more than they should. The suit was roundly condemned as an erosion of the sense of free will and personal responsibility in our society. Less widely remarked upon was that the teenagers were offering an accurate account of human behavior.

Consider the phenomenon of supersizing, wherein restaurant patrons are offered the chance to increase the portion size of their meal for some small amount of money. This presents a curious problem for the concept of free will: The patrons have already made a calculation about the amount of money they are willing to pay in return for a particular amount of food; however, when the question is re-asked – not 'Would you pay $5.79 for this total amount of food?' but 'Would you pay an additional 30 cents for more french fries?' – patrons often say yes, despite having answered 'No,' moments before, to an economically identical question.

Supersizing is expressly designed to subvert conscious judgment, and it works. By reframing the question, fast-food companies have found ways to take advantages of weaknesses in our analytical apparatus – weaknesses that are being documented daily in behavioral economics and evolutionary psychology.

This matters for more than just fat teenagers. Our legal, political, and economic systems – the mechanisms that run modern society – all assume that people are uniformly capable of consciously modulating their behaviors. As a result, we regard decisions they make as being valid, as in elections, or we hold them responsible for actions they take, as in contract law or criminal trials. Then, in order to get around the fact that some people obviously *aren't* capable of consciously modulating their behavior, we carve out ad-hoc exemptions. In U.S. criminal law, a fifteen-year-old who commits a crime is treated differently from a sixteen-year-old. A crime committed in the heat of the moment is treated specially. Some otherwise illegal actions are not crimes when their perpetrator is judged mentally incapable, whether through developmental disabilities or other forms of legally defined insanity, and so on.

This theoretical divide – between the mass of people with a uniform amount of free will and a small set of exceptional individuals – has been broadly stable for centuries, in part because it is based on ignorance. As long as we are unable to locate any biological source of free will, treating the mass of people as if each of them had the same degree of control over their lives makes perfect sense; no wiser judgments are possible. However, that binary notion of free will is being eroded, as our understanding of the biological antecedents of behavior improves.

Consider laws governing convicted pedophiles. Concern

about their recidivism rate has led to the enactment of laws that restrict their freedom based on what they might do in the future, even though this expressly subverts the notion of free will in the judicial system. The formula here – heinousness of crime times the likelihood of a repeat offense – creates a new, noninsane class of criminals whose penalty is indexed to a perceived lack of self-control.

But pedophilia is not unique in its measurably high recidivism rate. Rapists also have higher-than-average rates of repeat offense. Similarly, thieves of all varieties are likelier to become repeat offenders if they have short time-horizons or poor impulse control. Will we keep more kinds of criminals constrained after their formal sentence has been served, as we become better able to measure the likely degree of control they have over their own future actions? How can we, if we are to preserve the idea of personal responsibility? How can we not, once we are able to quantify the risk?

Criminal law is just one area where our concept of free will is eroding. We know that men make more aggressive decisions after they have been shown pictures of attractive female faces. We know that women are more likely to commit adultery on days when they are fertile. We know that patients committing involuntary physical actions routinely report, in order to preserve their (incorrect) belief that they are in control, that they decided to undertake those actions. We know that people will drive across town to save $10 on a $50 appliance but not on a $25,000 car. We know that the design of the ballot affects a voter's choices. And we are still in the early days even of understanding these effects; as we do, it becomes progressively easier to design everything from sales strategies to drug compounds to take advantage of them.

Conscious self-modulation of behavior is a spectrum. But we have been treating it as a single property – you are either capable of free will, or you fall into an exceptional category – because we have been unable to identify, measure, or manipulate the various components that go into such self-modulation. Those days are now ending, and everyone from advertisers to political consultants increasingly understands, in voluminous biological detail, how to manipulate consciousness in ways that weaken our notion of free will.

In the coming decades, our social and political concept of free will, based as it is on ignorance of its mechanisms, will be destroyed by what we learn about the actual workings of the brain. We can wait for that collision and decide what to do at that point, or we can begin thinking through what sort of legal, political, and economic systems we will need in a world in which our old conception of free will has been rendered inoperable.

The Limits of Introspection

MAHZARIN R. BANAJI

Mahzarin R. Banaji is Richard Clarke Cabot Professor of Social Ethics in the Department of Psychology, and Carol K. Pforzheimer Professor at the Radcliffe Institute for Advanced Study at Harvard University.

Conscious awareness makes up a sliver of the stuff the mind does. But it's the only aspect of the mind we subjectively experience, and hence the only aspect we believe exists. The truth is that thoughts, feelings, and behavior operate largely without deliberation or conscious recognition. It is the routinized, automatic, classically conditioned, precompiled aspects of our thoughts and feelings that make up a large part of who we are. We don't know what motivates us, even though we are certain we know just why we do the things we do, choose as we do, take action as we do. We have no idea that our perceptions and judgments are incorrect (as measured objectively), because we aren't confronted with such evidence, precisely because it remains outside our conscious awareness. We don't recognize that our behavior is often at variance with our own intentions and aspirations. The limits on introspection create bounds on our ethical judgments, not just on how we view the physical world.

The physiologist Helmholtz's notion of unconscious inference refers to the automatic act of making sense of what is perceived – without asking anybody's permission to go beyond the percept. When an action is observed – a simple action, such as a person reaching into a pocket – an unconscious inference occurs. In context, the movement is assumed to produce a wallet, or a diaper, or a stethoscope, or a gun. Such inferences, when they occur in ordinary social interaction, rely on the social categories that people belong to. Why assume that a grandmother leaning over a child is going to pull out a gun when a diaper is likely? So ordinary are such inferences that it makes sense that they happen rapidly and unconsciously.

From the objective analysis of mental processes, a science barely over a hundred years old, we know a lot about the very stuff we cannot intuit. We know, for instance, that fearing what is different from oneself is common. We know that disliking what is not part of the dominant part of social hierarchies is common. Put these together and what results is a particularly strong preference that majority-group members in any culture have for their own over others. Such tendencies to prefer one's own and prefer what's dominant is natural in the sense that it is a part of our evolutionary heritage and reinforced through learning that emphasizes the 'goodness' of one's own country, religion, and race/ethnicity. To recognize that in the world today what constitutes one's own group is complex, and what it means to be on the side of the dominant is not so clear (let alone morally questionable), shows the massive disadvantage of working without introspective access to such proclivities.

The mind sciences have made it possible to look inward, into the universe we carry around in the three-pounder in our heads, and in so doing we've learned about the strong limits on our power of introspection and the moral consequences of such limits. The ability to be fair in judging others objectively, the ability to act in accord with intention, the ability to treat members of ingroups and outgroups equally, the likelihood of privileging those who come from dominant and subordinate groups – these are heavily compromised mental acts, and invisibly so.

The only way out is to explore and understand the mind, using verifiable methods and confronting the facts that are tumbling out about who we are, without flinching. As everybody's favorite biologist, Richard Dawkins, said some thirty years ago: 'Let us understand what our own selfish genes are up to, because we may then at least have a chance to upset their designs, something that no other species has ever aspired to do.' The advice comes in handy, as we discover the 'mind bugs' that allow us to lead ourselves astray by denying ourselves (our conscious selves) access to the origins of out most fundamental thoughts and feelings.

Emily Dickinson wrote these words in a letter to a mentor asking him to tell her whether she was a decent poet: 'The sailor cannot see the north, but knows the needle can.' Without clear introspective access, we are such sailors. But a fact of life in this century is that we have the needle – in fact, several needles, the ones from science being the most obvious. These needles point toward the next (perhaps last?) frontier: that of allowing us to understand not just our place among other planets, our place among other species, but the very core of our nature.

What We Know May Not Change Us

BARRY C. SMITH

Barry C. Smith is a philosopher at Birkbeck, University of London, and editor (with Ernest Lepore) of *The Oxford Handbook of Philosophy of Language*.

Human beings, like everything else, are part of the natural world. The natural world is all there is. But to say that everything that exists is just part of the one world of nature is not the same as saying that there is just one theory of nature that will describe and explain everything that there is. Reality may be composed of just one kind of stuff and properties of that stuff, but we need many different kinds of theories, at different levels of description, to account for everything there is.

Theories at these different levels may not be reduced one to another. What matters is that they be compatible with one another. The astronomy Newton gave us was a triumph over supernaturalism because it united the mechanics of the sublunary world with an account of the heavenly bodies. In a similar way, biology allowed us to advance from a time when we saw life in terms of an *élan vital*. Today, the biggest challenge is to explain our powers of thinking and imagination, our abilities to represent and report our thoughts – the very means by which we engage in scientific theorizing. The final triumph of the natural sciences over supernaturalism will be an account of the nature of conscious experience. The cognitive and brain sciences have done much to make that project clearer, but we are still a long way from a fully satisfying theory.

But even if we succeed in producing a theory of human thought and reason, of perception, of conscious mental life, compatible with other theories of the natural and biological world, will we relinquish our cherished commonsense conceptions of ourselves as human beings, as selves who know ourselves best, who deliberate and decide freely on what to do and how to live? There is much evidence that we won't. As humans, we conceive ourselves as centers of experience, self-knowing and free-willing agents. We see ourselves and others as acting on our beliefs, desires, hopes, and fears, and as being responsible for much that we do and all that we say. Even as results in neuroscience begin to show how automated, routinized, and preconscious much of our behavior is, we remain unable to let go of the self-beliefs that govern our day-to-day rationalizing and our dealings with others.

We are perhaps incapable of treating others as mere machines, even if that turns out to be what we ourselves are. Our self-conceptions are firmly in place and sustained in spite of our best findings, and it may be a fact about human beings that this will always be so. We are curious and interested in neuroscientists' findings, and we wonder at them and about their applications to ourselves, but as the great naturalistic philosopher David Hume knew, nature is too strong in us, and it will not let us give up our cherished and familiar ways of thinking for long. Hume knew that however curious an idea and vision of ourselves we entertained in our study, or in the

BARRY C. SMITH   275

lab, when we returned to the world to dine and make merry with our friends our most natural beliefs and habits returned and banished our stranger thoughts and doubts. It is likely that whatever we have learned and whatever we know about the error of our thinking and about the fictions we maintain, they will remain the dominant guiding force in our everyday lives. We may not be comforted by this, but as creatures with minds who know they have minds – perhaps the only minded creatures in nature in this position – we are at least able to understand our own predicament.

Telling More Than We Can Know

RICHARD E. NISBETT

Richard E. Nisbett is a professor of psychology and codirector of the
Culture and Cognition Program at the University of Michigan. He is the
author of *The Geography of Thought: How Asians and Westerners Think
Differently . . . and Why*.

Do you know why you hired your most recent employee over
the runner-up? Do you know why you bought your last pair of
pajamas? Do you know what makes you happy and unhappy?

Don't be too sure. The most important thing that social psychologists have discovered over the last fifty years is that people
are unreliable informants about why they behaved as they did,
made the judgment they did, or liked or disliked something. In
short, we don't know nearly as much about what goes on in our
heads as we think. In fact, for a shocking range of things, we
don't know the answer to 'Why did I . . .?' any better than an
observer would.

The first inkling that social psychologists had about just how
ignorant we are about our thinking processes came from the
study of cognitive dissonance, beginning in the late 1950s.
When our behavior is insufficiently justified, we move our
beliefs into line with the behavior, so as to avoid the cognitive

dissonance we would otherwise experience. But we are usually quite unaware that we have done that, and when it is pointed out to us we recruit phantom reasons for the change in attitude.

Beginning in the mid 1960s, social psychologists started doing experiments about the causal attributions people make for their own behavior. If you give people electric shocks but tell them that you have given them a pill that will produce the arousal symptoms that are actually created by the shock, they will take much more shock than subjects without the pill. They have attributed their arousal to the pill and are therefore willing to take more shock. But if you ask them why they took so much shock they are likely to say something like 'I used to work with electrical gadgets and I got a lot of shocks, so I guess I got used to it.'

In the 1970s, social psychologists began asking whether people could be accurate about why they make simple judgments and decisions – such as why they like a certain article of clothing or a certain person. For example, in one study, experimenters videotaped a Belgian responding in one of two modes to questions about his philosophy as a teacher: He came across either as an ogre or a saint. They then showed the experimental subjects one of the two tapes and asked them how much they liked the teacher. Furthermore, they asked some of them whether the teacher's accent had influenced how much they liked him, and they asked others whether how much they liked the teacher influenced how much they liked his accent. Subjects who saw the ogre naturally disliked him a great deal, and they were quite sure that his grating accent was one of the reasons. Subjects who saw the saint realized that one of the reasons they were so fond of him was his charming accent. Subjects who were asked if their liking for the teacher could have influenced their judgment of his accent were insulted by the question.

Does familiarity breed contempt? On the contrary, it breeds liking. In the 1980s, social psychologists began showing people such stimuli as Turkish words and Chinese ideographs and asking them how much they liked them. They would show a given stimulus somewhere between one and twenty-five times. The more the subjects saw the stimulus, the more they liked it. Needless to say, the subjects did not find it plausible that the mere number of times they had seen a stimulus could have affected their liking for it. (You're probably wondering if laboratory rats are susceptible to the familiarity effect. The study has been done. Rats brought up listening to the music of Mozart prefer to move to the side of the cage that trips a switch allowing them to listen to Mozart rather than Schoenberg. Rats raised on Schoenberg prefer to be on the Schoenberg side. The rats were not asked the reasons for their musical preferences.)

Does it matter that we often don't know what goes on in our heads and yet believe that we do? Well, for starters, it means that we often can't answer accurately crucial questions about what makes us happy and what makes us unhappy. A social psychologist asked Harvard women to keep a daily record for two months of their mood states and also to record a number of potentially relevant factors in their lives, including amount of sleep the night before, the weather, general state of health, sexual activity, and day of the week (Monday blues? TGIF?). At the end of the period, subjects were asked to tell the experimenters how much each of these factors tended to influence their mood over the two-month period. The results? The women's reports of what influenced their moods were uncorrelated with what they had reported on a daily basis. If a woman thought her sexual activity had a big effect, a check of her daily reports was just as likely to show that it had no effect as to show that it did. To clinch the point, the subjects were asked to report on what influenced the moods of someone they didn't know: The degree of accuracy was just as great when a woman was rated by a stranger as when the woman rated herself!

If we were to think really hard about our reasons for behavior and preferences, might we be likely to come to the right conclusions? Actually, just the opposite may often be the case. A social psychologist asked people to choose which of several art posters they liked best. Some people were asked to analyze why they liked or disliked the various posters, and some were not asked, and everyone was given their favorite poster to take home. Two weeks later, the psychologist called people up and asked them how much they liked the art poster they had chosen. Those who had not analyzed their reasons liked their posters better than those who did.

It's certainly scary to think that we are ignorant of so much of what goes on in our heads, though we are almost surely better off taking with a large quantity of salt what we and others say about motives and reasons. Skepticism about our ability to read our own minds is safer than certainty that we can.

Still, the idea that we have little access to the workings of our minds is a dangerous one. The theories of Copernicus and Darwin were dangerous because they threatened, respectively, religious conceptions of the centrality of humans in the cosmos and the divinity of humans. Social psychologists are threatening a core conviction of the Enlightenment – that humans are perfectible through the exercise of reason. If reason cannot be counted on to reveal the causes of our beliefs, behaviors, and preferences, then the idea of human perfectibility is to that degree diminished.

The Quick-Thinking Zombies Inside Us

ANDY CLARK

Andy Clark holds the Chair in Logic and Metaphysics at the University of Edinburgh. He is the author of *Natural-Born Cyborgs: Minds, Technologies, and the Future of Human Intelligence*.

So much of what we do, feel, think, and choose is determined by unconscious, automatic uptake of cues and information.

Of course, advertisers will say they have known this all along. But only in recent years, with seminal studies by Tanya Chartrand, John Bargh, and others, has the true scale of our daily automatism really begun to emerge. Such studies show that it is possible (it is relatively easy) to activate racist stereotypes that impact our subsequent behavioral interactions – for example, yielding the judgment that your partner in a game or task is more hostile than an unprimed control would judge him/her to be. Such effects occur despite a subject's total and honest disavowal of those very stereotypes. In similar ways, it is possible to unconsciously prime you to feel older (and then you will walk more slowly).

In my favorite recent study, experimenters manipulate cues so that the subject forms an unconscious goal, whose (unnoticed) frustration results in the subject's losing confidence and performing worse at a subsequent task. The dangerous truth, it seems to me, is that these are not isolated laboratory events; instead, they reveal the massed woven fabric of our day-to-day existence. The underlying mechanisms impart an automatic drive toward the automation of all manner of choices and actions and don't discriminate between the trivial and the portentous.

It now seems clear that many of my major life and work decisions are made rapidly, often on the basis of ecologically sound but superficial cues, with slow deliberative reason busily engaged in justifying what the quick-thinking zombies inside me have already laid on the table. The good news is that without these mechanisms we would be unable to engage in fluid daily life or reason at all, and that very often they are right. The dangerous truth, though, is that we are indeed designed to cut conscious choice out of the picture whenever possible. This is not an issue about free will but simply about the extent to which conscious deliberation cranks the engine of behavior. Crank it it does, but not in anything like the way, or to the extent, that we may have thought. We had better come to grips with this before someone else does.

The Banality of Evil, the Banality of Heroism

PHILIP G. ZIMBARDO

Philip G. Zimbardo is professor emeritus of psychology at Stanford University and the author of, among other books, *Shyness: What It Is, What to Do About It.*

Those people who become perpetrators of evil deeds and those who become perpetrators of heroic deeds are basically alike, in being just ordinary, average people. The banality of evil is matched by the banality of heroism. Neither are the consequence of dispositional tendencies, nor are they special inner attributes of pathology or goodness residing within the human psyche or the human genome. Both emerge in particular situations at particular times, when situational forces play a compelling role in moving individuals across the line from inaction to action.

There is a decisive moment when the individual is caught up in a vector of forces emanating from the behavioral context. Those forces combine to increase the probability of acting to harm others or acting to help others. That decision may not be consciously planned or taken mindfully but impulsively driven by strong situational forces. Among them are group pressures and group identity, diffusion of responsibility, and a focus on the immediate moment without entertaining future cost or benefit.

The military-police guards who abused prisoners at Abu Ghraib, and the prison guards in my Stanford Prison Experiment who abused their prisoners, illustrate the temporary transition of ordinary individuals into perpetrators of evil, à la *Lord of the Flies*. I am not speaking here of those whose evil behavior is enduring and extensive – such tyrants as Idi Amin, Stalin, or Hitler. Nor of lifelong heroes. The heroic acts of Rosa Parks in a Southern bus, of Joe Darby in exposing the Abu Ghraib tortures, of New York City firefighters at the World Trade Center disaster, are acts of bravery at a specific time and place – whereas the heroism of Mother Teresa, Nelson Mandela, and Mahatma Gandhi consisted of valorous acts repeated over a lifetime. That chronic heroism is to acute heroism as valor is to bravery.

This view implies that any of us could as easily become heroes as perpetrators of evil, depending on how we are impacted by situational forces. We then want to discover how to limit, constrain, and prevent those situational and systemic forces that propel some of us toward social pathology.

It is equally important for our society to foster the heroic imagination in our citizens by conveying the message that anyone is a hero-in-waiting who will be counted on to do the right thing when the time comes to make the heroic decision.

Open-Source Currency

DOUGLAS RUSHKOFF

Douglas Rushkoff is a media analyst, documentary writer, and the author of Get Back in the Box: Innovation from the Inside Out.

It's not only dangerous and by most counts preposterous – it's happening. Open-source – or, in more common parlance, 'complementary' – currencies are collaboratively established units representing hours of labor that can be traded for goods or services in lieu of centralized currency. The advantage is that while the value of centralized currency is based on its scarcity, the bias of complementary or local currencies is toward their abundance.

So instead of having to involve the Fed in every transaction – and using money that requires being paid back with interest – we can invent our own currencies and create value with our labor. It's what the Japanese did at the height of the recession. No, not the Japanese government but unemployed Japanese people, who couldn't afford to pay health-care costs for their elderly relatives in distant cities. They created a currency through which someone could care for someone else's grandmother and accrue credits for someone else to take care of theirs.

Throughout most of history, complementary currencies existed alongside centralized currency. While local currency was used for labor and local transactions, centralized currencies were used for long-distance and foreign trade. Local currencies were based on a model of abundance – there was so much of it that people constantly invested it. That's why we saw so many cathedrals being built in the late Middle Ages and unparalleled levels of investment in infrastructure and maintenance. Centralized currency, on the other hand, needed to retain value over long distances and periods of time, so it was based on precious and scarce resources, such as gold.

The problem started during the Renaissance: As kings attempted to centralize their power, most local currencies were outlawed. This new monopoly on currency reduced entire economies into scarcity engines, encouraging competition over collaboration, protectionism over sharing, and fixed commodities over renewable resources. Today, money is lent into existence by the Fed or another central bank – and paid back with interest.

This cash is a medium, and like any medium it has certain biases. The money we use today is just one model of money. Turning currency into a collaborative phenomenon is the final frontier in the open-source movement. It's what would allow for an economic model that could support a renewable-energies industry, a way for companies such as Wal-Mart to add value to the communities it currently drains, and a way of working with money that doesn't have bankruptcy built in as a given circumstance.

Is the West Already on a Downhill Course?

DAVID BODANIS

David Bodanis is a writer and consultant and the author of *Passionate Minds: The Great Enlightenment Love Affair*.

I wonder sometimes whether the hyper-Islamicist critique of the West as a decadent force already on a downhill course might be true. At first it seems impossible: no country is richer than the United States, and no one has as powerful an army. Western Europe has vast wealth and university skills as well.

But what got me reflecting was the fact that in just four years after Pearl Harbor the United States had defeated two of the greatest military forces the world had ever seen: the German Army and the Imperial Japanese Navy. In that World War II period, everyone realized that there had to be restrictions on gasoline sales in order to preserve limited sources of gasoline and rubber. Profiteers were hated. But in the first four years after 9/11, Detroit automakers find it easy to continue paying off congressmen to ensure that gasoline-wasting SUVs aren't restricted in any way. American military forces have barely changed.

There are deep trends behind this. Technology is supposed to be speeding up, but if you think about it, airplanes have a similar feel and speed to those of thirty years ago; cars and oil rigs and credit cards and the operations of the New York Stock Exchange might be a bit more efficient than a few decades ago but also don't seem fundamentally different. Aside from the telephones, almost all the objects and daily habits in Steven Spielberg's twenty-five-year-old film *E. T.* are about the same as they are today.

What has transformed is the possibility of quick change; it's a lot harder than it was before. Patents for vague general ideas are much easier to get than they used to be, which slows down the introduction of new technology. Academics in biotech and other fields are wary about sharing their latest research with potential competitors, and that slows down the creation of new technology as well.

Moreover, there's a fear of falling from the increasingly fragile higher tiers of society, which means that social barriers are higher. I went to adequate but not extraordinary public schools in Chicago, but my children go to private schools. I suspect that many of my colleagues (unless they live in academic towns, where public schools are generally strong) are in a similar position. This is fine for our own children but not for those of the same potential who lack parents who can afford it.

Sheer inertia can mask such flaws for quite a while. The National Academy of Sciences has shown that, once again, the percentage of American-born university students studying the hard physical sciences has gone down. At one time, that didn't matter, for life in the United States – and at the top U.S. universities – was an overwhelming lure for ambitious youngsters from Seoul and Bangalore. They would come to America and make up the gap. But already there are signs of that slipping, and who knows if enough of those energetic foreign students will still be coming to America or Western Europe in another decade or two.

Another sort of inertia is coming to an end as well. The first generation of migrants from farm to city brought with them the attitudes of their farm world; the first generation of migrants from blue-collar city neighborhoods to upper-middle-class professional life bring similar attitudes of responsibility as well. They often vote against their short-term economic interests because it's 'the right thing to do'; they engage in philanthropy toward individuals from backgrounds very different from their own. But why? In many parts of America and Europe, the circumstances creating those attitudes no longer exist. When they finally melt away, will what replaces them be strong enough for us to survive?

Technology Can Untie the United States

JUAN ENRIQUEZ

Juan Enriquez, formerly the founding director of Harvard Business School's Life Sciences Project, is the CEO of Biotechonomy. He is the author of *The Untied States of America*.

Everyone grows and dies; the same is true of countries. The only question is how long one can postpone the inevitable. In the case of some countries, life spans can be very long, so it is worth asking whether the United States is in adolescence, middle age, or old age. Do science and technology accelerate or offset the demise? And finally, how many stars will be in the U.S. flag in fifty years?

There has yet to be a single U.S. president buried under the same flag he was born under, yet we often take continuity for granted. Just as almost no newlyweds expect to divorce, citizens rarely assume that their beloved country, flag, and anthem might end up an exhibit in an archeology museum. But countries rich and poor, Asian, African, and European have been untying time and again. In the last five decades, the number of United Nations members has tripled. This trend goes way beyond the decolonization of the 1960s, and it is not exclusive to failed states; it is a daily debate within the United Kingdom,

Italy, France, Belgium, the Netherlands, Austria, and many other nations.

So far, the Americas have remained mostly impervious to this global trend, but even if in God you trust, there are no guarantees. Over the next decade, waves of technology will wash over the United States. Almost any applied field you care to look at promises extraordinary change, opportunities, and challenges. (Witness the entries in this book.) How countries adapt to massive and rapid upheaval will go a long way toward determining the eventual outcomes. To paraphrase Darwin, it is not the strongest, not the largest, that survive; rather it is those best prepared to cope with change.

It is easy to argue that the United States could be a larger, more powerful country in fifty years. But it is also possible that like so many other great powers, it could begin to unravel. This is not something that depends on what we decide to do fifty years hence. To a great extent it depends on what we choose to do, or choose to ignore, today. There are more than a few worrisome trends.

Future ability to generate wealth depends on technoliteracy. But educational excellence, particularly in grammar schools and high schools, is far from uniform, and it is not world-class. Time and again, the United States does poorly, particularly in regard to math and science, when compared with its major trading partners. Internally there are enormous disparities among schools and among the number of students that pass state competency exams and what federal tests tell us about the same students. There are also large gaps in technoliteracy among ethnic groups. By 2050, close to 40 percent of the U.S. population will be Hispanic and African American. These groups receive 3 percent of the PhDs in math and science today.

How we prepare kids for a life-sciences-, materials-, robotics-, IT-, and nanotechnology-driven world is critical, but the federal government currently invests $22,000 in those over sixty-five and just over $2,000 in those under sixteen.

As ethnic, age, and regional gaps in the ability to adapt increase, many become wary and frustrated by open borders, free trade, and smart immigrants. Historically, when others use newfangled ways to leap ahead, it can lead to a conservative response. This is likeliest within those societies and groups that have the most to lose, often those who have been the most successful. There is frequently a reflexive response: 'Stop the train; I want to get off.' Or, as the Red Sox have said, 'Just wait till last year!' No more teaching evolution, no more research into stem cells, no more Indian or Chinese or Mexican immigrants, no matter how smart or hardworking they might be. These individual battles are signs of a creeping xenophobia, isolationism, and fury.

Within the United States, there are many who are adapting successfully. They tend to concentrate in a very few zip codes – life-science clusters like 92121 (between Salk, Scripps, and UC San Diego) and techno-empires like 02139 (MIT). Most of the nation's wealth and taxes are generated by a few states – and inside these states, within a few square miles. Those who live in these areas are the most affronted by restrictions on research, the lack of science-literate teenagers, and the reliance on God instead of science.

Politicians well understand these divides, and they have gerrymandered their districts to reflect them. Because competitive congressional elections are rarer today than turnovers within the Soviet Politburo, there is hardly ever an open discussion as to why other parts of the country act and think so differently. The

Internet and cable TV further narrowcast news and views, tending to reinforce what one's neighbors and communities already believe. Positions harden. Anger at 'the others' mounts.

Add a large and increasing debt to this equation, along with politicized religion, and the mixture becomes explosive. The average household now owes over $88,000 and the present value of what we have promised to pay is now about $473,000. There is little willingness within Washington to address a growing deficit, never mind the current account imbalance. Facing the next electoral challenge, few seem to remember that the last act of many an empire is to drive itself into bankruptcy.

Sooner or later, we could witness some bitter arguments about who gets and who pays. In developed country after developed country, it is often the richest, not the ethnically or religiously repressed, that first seek autonomy and eventually dissolution. In this context it is worth recalling that New England, not the South, has been the most secession-prone region. As the country expanded, New Englanders attempted to include the right to untie into the Constitution; the argument was that as this great country expanded south and west, they would lose control over their political and economic destiny. Perhaps this is what led to four separate attempts to untie the Union.

When we assume stability and continuity, we can wake up to irreconcilable differences. Science and a knowledge-driven economy can allow a few folks to build powerful and successful countries very quickly – witness Korea, Taiwan, Singapore, Ireland – but changes of this magnitude can also bury or split the formerly great who refuse to adapt, as well as those who practice bad governance. If we do not begin to address some current divides quickly, we could live to see an Un-Tied States of America.

Democracy May Be On Its Way Out

HAIM HARARI

HAIM HARARI is a theoretical physicist and former president of the Weizmann Institute of Science.

Democracy may be on its way out. Future historians may determine that democracy will have been a one-century episode. It will disappear. This is a sad, truly dangerous, but very realistic idea (or, rather, prediction).

Falling boundaries between countries, cross-border commerce, merging economies, instant global flow of information, and numerous other features of our modern society all lead to multinational structures. If you extrapolate this irreversible trend, you get the entire planet becoming one political unit. But in this unit, antidemocracy forces are now a clear majority. This majority increases by the day, due to demographic patterns. All democratic nations have slow, vanishing, or negative population growth, while all antidemocratic and uneducated societies multiply fast. Within democratic countries, most well-educated families remain small, while the least educated families are growing fast. This means, both at the individual level and at the national level, that the more people you represent, the less economic power you have. In a knowledge-based economy, in which the number of working hands is less important, this situation is much more nondemocratic than in the industrial age. As long as the upward mobility of individuals and nations could neutralize the phenomenon, democracy was tenable. But when we apply this analysis to the entire planet as it evolves now, we see that democracy may be doomed.

To this idea we must add the regrettable fact that authoritarian multinational corporations, by and large, are better managed than democratic nation states. Religious preaching, TV sound bites, cross-boundary TV incitement, and the freedom to spread rumors and lies through the Internet all abet brainwashing and lack of rational thinking. Proportionately, more young women are growing up in societies that discriminate against them than in the more egalitarian societies, increasing the worldwide percentage of women who are treated as second-class citizens. Educational systems in most advanced countries are in a deep crisis, while modern education in many developing countries is almost nonexistent. A small, well-educated technological elite is becoming the main owner of intellectual property, which is, by far, the most valuable economic asset, while the rest of the world drifts toward fanaticism of one kind or another. In sum, the unavoidable conclusion is that democracy, our least bad system of government, is on its way out.

Can we invent a better system? Perhaps. But this cannot happen if we are not allowed to utter the sentence 'There may be a political system that is better than democracy.' Today's political correctness does not allow one to say such things. The result of this prohibition will be an inevitable return to some kind of totalitarian rule – different from that of the emperors, the colonialists, or the landlords of the past, but not more just.

Alternatively, open and honest thinking about this issue may lead either to a worldwide revolution in educating the poor masses, thus saving democracy, or to a careful search for a just (repeat, just) and better system.

Marx Was Right: The State Will Evaporate

JAMES O'DONNELL

James O'Donnell is a classicist and cultural historian and provost of Georgetown University. He is the author of *Avatars of the Word: From Papyrus to Cyberspace* and *Augustine: A New Biography*.

From the earliest Babylonian and Chinese civilizations, we have agreed that human affairs depend on an organizing power in the hands of a few people (usually with religious charisma to undergird their authority) who reside in a functionally central location. 'Political science' assumes, in its etymology, the *polis*, or city-state of Greece, as the model for community and government.

It is remarkable how little of human excellence and achievement has ever taken place in capital cities and around those elites, whose cultural history is one of self-mockery and implicit acceptance of the marginalization of the powerful. Borderlands and frontiers (and even suburbs) are where the action is.

As long as technologies of transportation and weaponry emphasized geographic centralization and concentration of forces, the general or emperor or president in his capital with armies at his beck and call was the most obvious focus of power. Enlightened government constructed mechanisms to restrain and channel such centralized authority but did not effectively challenge it.

So what advantage is there today to the nation state? Boundaries between states enshrine and exacerbate inequalities and prevent the free movement of peoples. Large and prosperous state and state-related organizations and locations attract the envy and hostility of others and are sitting-duck targets for terrorist action. Technologies of communication and transportation now make geographically defined communities increasingly irrelevant and provide the new elites and new entrepreneurs with ample opportunity to stand outside them. Economies construct themselves in spite of state management, and money flees taxation as relentlessly as water follows gravity.

Who will undergo the greatest destabilization as the state evaporates and its artificial protections and obstacles disappear? The sooner all that happens, the more likely it is to be the United States. The longer it takes . . . well, perhaps the new Chinese Empire isn't quite the landscape-dominating leviathan of the future that it wants to be. Perhaps in the end it will be Mao who was right, and a hundred flowers will bloom there.

Following Sisyphus

HOWARD GARDNER

Howard Gardner is the John H. and Elisabeth A. Hobbs Professor in Cognition and Education at the Harvard Graduate School of Education and adjunct professor of psychology at Harvard University. Among his most recent books are *Good Work: When Excellence and Ethics Meet* (2001), *Changing Minds* (2004), and *Multiple Intelligences: New Horizons* (2006).

According to myth, Pandora unleashed all evils upon the world; only hope remained inside the box. Hope for human survival and progress rests on two assumptions: (1) Human constructive tendencies can counter human destructive tendencies, and (2) human beings can act on the basis of long-term considerations, rather than merely short-term needs and desires. My personal optimism, and my years of research on 'good work,' could not be sustained without these assumptions.

Yet I lie awake at night with the dangerous thought that pessimists may be right. For the first time in history (as far as we know), we humans live in a world we could completely destroy. The human destructive tendencies described in the past by Thomas Hobbes and Sigmund Freud and the 'realist' picture of human beings embraced more recently by many sociobiologists, evolutionary psychologists, and game theorists might be correct; these tendencies could overwhelm any proclivities toward altruism, protection of the environment, control of weapons of destruction, progress in human relations, or seeking to become good ancestors. As one vivid data point: There are few signs that the unprecedented power possessed by the United States is being harnessed to positive ends.

Strictly speaking, what will happen to the species or the planet is not a question for scientific study or prediction; it is a question of probabilities, based on historical and cultural considerations as well as on our most accurate description of human nature(s). Yet science has recently invaded this territory, with its assertions of a biologically based human moral sense. Those who assert a human moral sense are wagering that in the end human beings will do the right thing. Of course, human beings have the ability to make moral judgments – that is a mere truism. But my dangerous thought is that this moral sense is up for grabs, that it can be mobilized for destructive ends (one society's terrorist is another society's freedom fighter) or overwhelmed by other senses and other motivations, such as the quest for power, instant gratification, or the annihilation of one's enemies.

I will continue to do what I can to encourage good work – in that sense, Pandoran hope remains. But I will not look to science, technology, or religion to preserve life. Instead, I will follow Albert Camus's injunction, in his portrayal of another mythic figure endlessly attempting to push a rock up a hill: One should imagine Sisyphus happy.

How Can I Trust, in the Face of So Many Unknowables?

ERNST PÖPPEL

Ernst Pöppel is a neuroscientist, chairman of the board of directors of the Human Science Center, and chair of the Institute of Medical Psychology, Ludwig-Maximilians-Universität, Munich, and the author of *Mindworks: Time and Conscious Experience.*

The average life expectancy of a species on this globe is just a few million years. From an external point of view, it would be nothing special if humankind were to suddenly disappear. We have been here for some time. With humans no longer around, evolutionary processes would have an even better chance to fill in all those ecological niches that have been created by human activities. As we change the world, and as thousands of species are lost every year because of human activities, we provide a new and productive environment for the creation of new species. Thus, humankind is creative with respect to providing a frame for new evolutionary trajectories – and would be even more creative if it disappeared altogether. If somebody (unfortunately not our descendant) were to visit this globe sometime later, he would meet many new species that owe their existence to the presence and disappearance of humankind.

But this is not going to happen, because we are doing science. With science we apparently get a better understanding of basic principles in nature, we have a chance to improve quality of life, and we can develop means to extend the life expectancy of our species. However, some of these scientific activities have a paradoxical effect, in that they may result in a higher risk of our disappearance. Maybe science will not be so effective after all in preventing it.

Now comes my dangerous idea. My own personal dangerous idea is my belief in science.

In all my research in the field of temporal perception and visual processes, I have had a basic trust in the science and I believe the results I have obtained. And I believe the results of others. But why? I know that there are many unknown and unknowable variables that are part of the experimental setup and cannot be controlled. How can I trust, in the face of so many unknowables? Furthermore, can I really rely on my thinking? Can I trust my own eyes and ears? Can I be so sure about my scientific work that I communicate the results with pride to others?

If I look at the complexity of the brain, how is it possible that something reasonable comes out of this network? How is it possible that a face I see or a thought of mine can maintain their identity over time? If I have no access to what goes on in my brain, how can I be proud, (how can anybody be proud) of scientific achievements?

A Twenty-Four-Hour Period of Absolute Solitude

LEO M. CHALUPA

Leo M. Chalupa is Distinguished Professor of Ophthalmology and
Neurobiology at the University of California, Davis.

Our brains are constantly subjected to the demands of multi-
tasking and a seemingly endless cacophony of information –
from cellphones, e-mails, computers, and cable television, not
to mention such archaic venues as books, newspapers, and mag-
azines.

   This induces an unrelenting barrage of neuronal activity that
in turn produces long-lasting structural modification in virtually all compartments of the nervous system. A fledging
industry touts the virtues of exercising your brain for self-
improvement. Programs are offered for how to make virtually
any region of your neocortex a more efficient processor. Parents
are urged to begin such regimes in preschool children, and
adults are told to take advantage of their brain's plastic properties for professional advancement. The jury is out on the
efficacy of such claims, but one thing is clear: Even if brain exercise works, the subsequent waves of neuronal activity stemming
from simply living a modern lifestyle are likely to eradicate its
hard-earned benefits.

My dangerous idea is that what's needed to attain optimal brain performance – with or without brain exercise – is a twenty-four-hour period of absolute solitude. By absolute solitude I mean no verbal interactions of any kind – written or spoken, live or recorded – with another human being. I would venture that a significantly greater number of people reading these words have tried skydiving than have experienced one day of absolute solitude.

What to do to fill the waking hours of such a day? That's a question each person would have to answer for himself or herself. Unless you've spent time in a monastery or in solitary confinement, it's unlikely that you've had to deal with this issue. The only activity not proscribed is thinking. Imagine if everyone in this country had the opportunity to do nothing but engage in uninterrupted thought for one full day a year!

A national day of absolute solitude would do more to improve the brains of Americans than any other one-day program. (I leave it to the lawmakers to figure out how to implement this proposal.) The danger inherent in the idea is that a day of uninterrupted thinking might cause irrevocable upheavals in much of what our society holds sacred. Whether that would improve our present state of affairs cannot be guaranteed.

Afterword

RICHARD DAWKINS

Dangerous ideas are what has driven humanity onward, usually to the consternation of the majority in any particular age who thrive on familiarity and fear change. Yesterday's dangerous idea is today's orthodoxy and tomorrow's cliché. Surely somebody must have said that? If not I'll have to say it myself, although only to pull back in a hurry. Such seductive generalizations conceal a dangerous asymmetry. Although it is true that hindsight can recognize accepted norms that were once dangerous ideas, it is also true that most dangerous ideas from the past neither deserved nor received eventual acceptance. It is not enough for an idea to be dangerous. It must also be good.

Scientists pay lip service to the view that an idea must stand on its own merits, not on the authority of its inventor. There is no scientific *Führer*, pope, or prophet of whom we are tempted to say, '*X* is his idea so *X* must be right.' But scientists are only human, and we inevitably take note of a proven track record. If a star scientist whose ideas have worked in the past comes up with a new one, we prick up our ears. Especially if the new idea is a dangerous one.

Where scientists are concerned, John Brockman has the most enviable address book in America. His annual *Edge* Question yields a book whose table of contents on its own is well worth reading. Here is a set of authors with something to say, and with outstanding credentials to say it, all faced with the same seemingly simple question – in this case, 'What is your dangerous idea?' What answers will the *Edge* circle come up with? What surprising meanings, indeed, will they discover for the question? Dangerous to whom? Or to what? The 108 contributors to this book ply the spectrum. There's danger to the world or to the future of humanity and life. There's danger to vested interests whose *amour propre* might be threatened. There's danger to one's own personal peace of mind or sense of cosmic worth. There's danger in the sense of ideas that are intellectually daring or bold – that push the envelope, to employ the fashionable cliché – which doesn't necessarily imply danger in any of the other ways. Happily, in modern America there is no need to talk about ideas that threaten the thinker's life because they are deemed unacceptable by the prevailing society. Galileo was prevented, on pain of physical harm, from publishing his dangerous ideas. Darwin was more fortunate in his time, although he arguably censored his dangerous idea for two decades for fear of upsetting his wife, and the society of which she was a part. Closer to our own time, in Lysenko's Russia, ideas that today's geneticists consider commonplace – indeed, simply true – could not be uttered without risking public humiliation and imprisonment.

This book presents to us 108 top intellectuals from the *Edge* on-line salon, famed for their good ideas (or, in one or two cases, notoriously bad ideas). What, then, are their *dangerous* ideas, and are they any good? I found that I could analyze the answers as a kind of poll. How many opt for doom and foreboding – global warming, terrorist meltdown, and similar apocalyptic jeremaiads? By my count, 11, although some of these were anti-Jeremiahs whose dangerous idea is that the dangers are exaggerated. I counted 24 whose dangerous idea concerns society, 20 whose dangerous idea touches on psychology, and 14 on politics or economics. Eleven chose topics that, in one way or another, concern religion, broadly defined. Six explore the cosmic angst that seems to follow from, for example, the belief that we are alone in the universe, or the belief that there is nobody at home in our skulls, nothing that could honestly answer to the name of 'soul.' Six authors take a self-referential approach to the *Edge* Question, discussing as a dangerous idea the very idea of asking for dangerous ideas – or, in one case, the very idea that ideas *can* be considered dangerous.

Those tallies are not mutually exclusive. I did, however, recognize one exclusive pair of categories, and I forced myself to place every contribution in one or the other of them. It seemed to me that there is a nonoverlapping and exhaustive distinction between ideas that are false or true about the real world (factual matters, in the broad sense) and ideas about what we ought to do – normative or moral ideas, for which the words 'true' and 'false' have no meaning. It is perhaps unsurprising that a group predominantly made up of scientists should favor 'is' (factual, true-or-false) ideas over 'ought' (normative, policy) ideas, but not by a great margin. I make it 68 factual to 40 normative ideas.

Are there any dangerous ideas that are conspicuously under-represented in this book? I have two suggestions, both of which can be spun into either the 'is' or the 'ought' box. First, I noticed only fleeting references to eugenics, and they were disparaging. In the 1920s and '30s, scientists from both the political left and right would not have found the idea of designer babies particularly dangerous – though of course they would not have used that phrase. Today I suspect that the idea is too dangerous for comfortable discussion, even under the license granted by a book like this, and my conjecture is that Adolf Hitler is responsible for the change. Nobody wants to be caught agreeing with that monster, even in a single particular. The spectre of Hitler has led some scientists to stray from 'ought' to 'is' and deny that breeding for human qualities is even possible. But if you can breed cattle for milk yield, horses for running speed, and dogs for herding skill, why on earth should it be impossible to breed humans for mathematical, musical, or athletic ability? Objections such as 'These are not one-dimensional abilities' apply equally to cows, horses, and dogs and never stopped anybody in practice.

I wonder whether, some sixty years after Hitler's death, we might at least venture to *ask* what the moral difference is between breeding for musical ability and forcing a child to take music lessons. Or why it is acceptable to train fast runners and high jumpers but not to breed them. I can think of some answers, and they are good ones, which would probably end up persuading me. But hasn't the time come when we should stop being frightened even to put the question?

My other surprise omission from this list of dangerous ideas concerns the unspoken assumption of human moral uniqueness. It is harder than most people realize to justify the unique and exclusive status that *Homo sapiens* enjoys in our unconscious assumptions. Why does 'pro life' always mean 'pro *human* life?' Why are so many people outraged at the idea of killing an eight-celled human conceptus while cheerfully masticating a steak that cost the life of an adult, sentient, and probably terrified cow? What precisely is the moral difference between our ancestors' attitude toward slaves and our attitude toward nonhuman animals? Probably there are good answers to these questions. But shouldn't the questions themselves at least be *put?*

One way to dramatize the nontriviality of such questions is to invoke the fact of evolution. We are connected to all other species continuously and gradually, via the ancestors we share with them. But for the historical accident of extinction, we would be linked to chimpanzees via an unbroken chain of happily interbreeding intermediates. What would – should – be the moral and political response of our society if relict populations of all the evolutionary intermediates were now discovered in Africa? What should be our moral and political response to future scientists who use the completed human and chimpanzee genomes to engineer a continuous chain of living, breathing, and mating intermediates, each capable of breeding with its nearer neighbors in the chain, thereby linking humans to chimpanzees via a living cline of fertile interbreeding.

I can think of formidable objections to such experimental breaches of the wall of separation around *Homo sapiens*. But at the same time, I can imagine benefits to our moral and political attitudes that might outweigh the objections. We know that such a living daisy chain is in principle possible, because all the intermediates have lived – in the chain leading back from ourselves to the common ancestor with chimpanzees and then the chain leading forward from the common ancestor to chimpanzees. It is therefore a dangerous but not too surprising idea that one day the chain will be reconstructed – a candidate for the 'factual' box of dangerous ideas. And – moving across to the 'ought' box – mightn't a good moral case be made that it *should* be reconstructed? Whatever the undoubted moral drawbacks of such a project, it would at least jolt humanity finally out of the absolutist and essentialist mindset that has so long afflicted us.

INDEX